Bob Gough, PhD

An Encyclopedia
of Small Fruit

Pre-publication
REVIEW

"**G**ets top votes from me. Any fruit garden enthusiast, novice or experienced, would enrich their library with this reference. . . . a valuable reference. . . . I know I'll be using this book as a resource in my office, and I'd highly recommend it to anyone with an interest in culturing small fruits."

Amy P. K. Grandpre
*Applied Associate Degree
in Nursery-Horticulture Management,
Horticulture Assistant,
Yellowstone County Extension*

AN ENCYCLOPEDIA
of SMALL FRUIT

AN ENCYCLOPEDIA
of SMALL FRUIT

BOB GOUGH, PHD

CRC Press
Taylor & Francis Group
Boca Raton   London   New York

CRC Press is an imprint of the
Taylor & Francis Group, an **informa** business

CRC Press
Taylor & Francis Group
6000 Broken Sound Parkway NW, Suite 300
Boca Raton, FL 33487-2742

© 2008 by Taylor & Francis Group, LLC
CRC Press is an imprint of Taylor & Francis Group, an Informa business

No claim to original U.S. Government works
Printed in the United States of America on acid-free paper
10 9 8 7 6 5 4 3 2 1

International Standard Book Number-13: 978-1-56022-939-1 (Softcover)

Library of Congress Cataloging-in-Publication Data

Gough, Bob.
    An encyclopedia of small fruit / Bob Gough.
        p. cm.
    ISBN: 978-1-56022-938-4 (hard : alk. paper)
    ISBN: 978-1-56022-939-1 (soft : alk. paper)
    1. Fruit-culture—Encyclopedias. I. Title.

SB354.4.G68 2007
634.03—dc22   2007034355

Visit the Taylor & Francis Web site at
http://www.taylorandfrancis.com

and the CRC Press Web site at
http://www.crcpress.com

CONTENTS

ABOUT THE AUTHOR

Robert E. Gough, PhD, is Associate Dean for Academic Programs in the College of Agriculture at Montana State University. Dr. Gough is the author, editor, or co-editor of numerous book chapters, over 350 articles, and 14 books, including *The Highbush Blueberry and Its Management; Small Fruits in the Home Garden; Glossary of Vital Terms for the Home Gardener;* and *Blueberries* (all from Haworth). He is a current or former member of many professional and honorary societies, including the International Society for Horticultural Science (ISHS), the American Pomological Society (APS), the American Society for Horticultural Science (ASHS), and the American Association for the Advancement of Science (AAAS). In addition to being the recipient of several professional awards and honors, Dr. Gough's newspaper columns, television segments, and radio programs regularly appear throughout the Montana, South Dakota, and Wyoming area.

Introduction

Book introductions usually are an author's attempt to tell what a book is about and how it should be used, but let me begin this introduction by telling what this book is not about. It is not about providing a complete and in-depth review of all small fruits and their culture, nor is it a grower's manual for commercial production. Rather, this book is an overview of temperate and tropical small fruits with notes on their culture. Some entries provide far more information than others simply because far more is known about the culture of some fruits than of others.

This is a book for the home grower and general fruit garden enthusiast. The text is written in nontechnical language whenever possible and provides information on all aspects of fruit culture that the home grower would require. A deficiency in specific information related, say, to pollination or fertilization of a species generally means that the literature is deficient in that information.

Main entries of common names appear in bold letters. Where an entry is not a standard name for a species, or where several common names are in use, the reader is directed to a specific entry. Following the main entry is the name of the family to which the species belongs. This in turn is followed by all synonyms in small capital letters. If the name appearing in the main entry applies to several species, such as in the case of "bearberry," the more common of those species are listed and described under the heading. In nearly all cases the scientific nomenclature adheres to that of *Hortus Third: A Concise Dictionary of Plants Cultivated in the United States and Canada*. Within each heading, species are described generally in simple language and directions for culture are given in systematic arrangement. Also provided in addition to cultural practices is a general history of the genus and the use of its fruit, notes on propagation, production information if it is available, and the hardiness zone(s) to which the species is adapted. The hardiness zones are those given by the U.S. Department of Agriculture. Finally, I have included for the home grower a brief glossary to explain some of the more complicated terms used in the book.

At times I have made an arbitrary distinction between a small (bush) fruit and a tree fruit. For example, chokecherry and American plum can be pruned to a single trunk and called tree fruit. However, their natural habit of growth, wherein they sucker profusely and form thickets, renders them bush fruit, and so they are included here.

Common names are sometimes confusing, with toxic and non-toxic species sharing the same name, as in the nightshades. Never eat fruit of a plant you cannot positively identify. Further, some people may be allergic to fruit commonly considered edible. Be careful what you eat.

Last, this book is the product of work over the past several years. Many have contributed to technical aspects of its formatting and entries. However, its completion at this time is the result of constant urging, encouragement, technical help, and strong support from my wife, Cheryl, and it is to her that I dedicate this work.

A

acerola: *See* **cherry, Barbados.**

actinidia, bower *(Actinidia arguta): See* **kiwifruit.**

Akebia: Lardizabalaceae. This genus contains two species, both monoecious, twining, semievergreen shrubs or vines native to temperate areas of east Asia. Leaves alternate, palmately compound; leaflets three to five, stalked, emarginated; flowers purplish, in axillary racemes, sepals three; male flowers toward the apex of the raceme, stamens six; female flowers below, pistils three to twelve; fruit ovoid-oblong, blue or purple berry, dehiscent along one side; seeds many, black (Hortus, 1976). Plants of this genus do best in full sun with well drained soil and are propagated by seeds, hardwood and softwood cuttings, and root division.

Akebia quinata (Houtt.) Decne. FIVE-LEAF AKEBIA; CHOCOLATE VINE. Leaflets five, entire. The fragrant, dark-colored flowers open in spring. Native to Japan, China, and Korea. The showy fruit varies in size but is normally about 6 to 8 centimeters (2-3 inches) long and 5 centimeters (2 inches) in diameter, violet-purple, bloomy, with yellowish-green, homogeneous pulp containing 40 to 50 oblong seeds. Flavor sweetish, pleasant, but insipid (Hedrick, 1919; Rehder, 1947).

Akebia trifoliata (Thunb.) G. Koidz. [*A. lobata* Decne.]. THREE-LEAF AKEBIA. Leaflets three, coarsely toothed or entire. Native to Japan and China. Cultivated before 1890. This is less handsome than preceding species (Rehder, 1947).

alkekengi *(Physalis alkekengi): See* **cherry, ground.**

almond, flowering *(Prunus japonica): See* ***Prunus*.**

amatungula: *See* **Carissa.**

apricot, Manchurian bush *(Prunus armeniaca* var. *mandshurica): See* ***Prunus*.**

apricot, Siberian bush *(Prunus armeniaca* var. *siberica): See* ***Prunus*.**

apple, baked *(Rubus chamaemorus): See* **blackberry.**

apple, belle *(Passiflora laurifolia): See* **passionfruit.**

apple, conch *(Passiflora maliformis): See* **passionfruit.**

appleberry: Pittosporaceae. About eight species make up the genus *Billardiera.* All are small, twining shrubs native to Australia. Leaves alternate; flowers usually yellow or purple, solitary or clustered; fruit is a berry.

  Billardiera longiflora Labill (Hortus, 1976). Entire leaves ovate to linear and about 3 centimeters (1 inch) long; flowers greenish-yellow becoming purple. The fruit is blue or violet with a pleasant, subacid taste (Hedrick, 1919). These plants are grown in greenhouses or out-doors, being hardy to zone 9.

applerose: Rosaceae. WOLLY DOD'S ROSE. There are more than 100 species of plants in the genus *Rosa.* All are more or less prickly shrubs mostly native to temperate areas of the northern hemisphere. Leaves alternate, usually odd-pinnate; flowers solitary, with five pet-als and many stamens. The fruit of all species is a fleshy hip contain-ing hairy achenes (Hortus, 1976). The flowers of many hybrids are also used to flavor honey, confections, teas, salads, and sorbets (Markle et al., 1998).

  Rosa villosa L. (*R. mollis* Sm.; *R. pomifera* J. Herrm.). Erect shrub up to 2 meters (6 feet) in height with slender, subulate prickles; leaf-lets two to four pairs; flowers one to three in a cluster, semidouble, pink or nearly red, 2.5 to 5 centimeters (1-2 inches) diameter. Native from Europe to Iran and hardy to zone 6, the plants were first culti-vated in 1771 (Rehder, 1947). These plants have growing conditions similar to those for other roses and bear a hip with pleasant, subacid pulp that is eaten out of hand, used to make beverages, or made into conserves (Hendrick, 1919). Plants can be propagated by seed ex-tracted from ripe, presenescent fruit, stratified immediately at 2 to 4°C (35.6-39.2°F) for several months and planted in seedbeds in early spring. Where winters are severe, make hardwood cuttings in early winter, store them in damp peat moss at 4°C (39°F) until spring, then plant in rows outdoors. In areas with mild winters take cuttings and field-plant immediately in early fall. Softwood cuttings root readily under mist or in high humidity chambers (Hartmann et al., 2002).

ash, mountain: Rosaceae. SERVICEBERRY. The genus *Sorbus* contains about 85 species of deciduous trees and shrubs native to the northern hemisphere. Leaves alternate, simple; flowers white, mostly perfect and borne in terminal clusters. Most species are hardy to the north and grown as ornamentals, although the fruit of many is used in preserves or in the manufacture of spirits (Hortus, 1976; Hedrick, 1919). The fruit is eaten fresh after being exposed to a frost or when overripe, as medlar is consumed. Also, fruit may be fermented into wine, dried like prunes, or made into jelly. The bark has been used in the tanning industry.

Sorbus americana Marsh. AMERICAN MOUNTAIN ASH, DOG-BERRY, MISSEY-MOOSEY, ROUNDWOOD. This shrubby tree bears bright red fruit up to 0.5 centimeters (0.25 inches) in diameter. It is hardy to zone 2 and a native to eastern North America. This species may be propagated by autumn-sown seeds after removal of the pulp, or by layering (Hartmann et al., 2002).

avellana *(Corylus avellana): See* **filbert.**

banana, custard: Annonaceae. The genus *Asimina* contains about eight species native to North America. Leaves alternate, simple; white to purple flowers axillary, bisexual; fruit an ellipsoid to oblong berry.

Asimina triloba (L.) Dunal. BANANA TREE, HOOSIER BANANA, INDIANA BANANA, MICHIGAN BANANA, NEBRASKA BANANA, PAWPAW, POOR MAN'S BANANA, SMALL FLOWER. This is the most commonly cultivated species of *Asimina,* hardy to zone 5, and found from New York south to Florida and west to Texas. It forms a large bush or sometimes a small tree and is the only member of this genus grown in the temperate zone. As of the early 1990s only one planting was in commercial production in the United States (Markle et al., 1998), with most of the crop being collected from the wild. Italy and Japan also have limited commercial production, but the fruit has enjoyed local popularity for years. Recently the bark has been found to contain highly effective pesticidal and anticancer compounds (Alkofahi et al., 1989; Ruppercht et al., 1986). The number of cultivars available is limited, with 'Davis', 'Overleese', and 'Sunflower' the most popular. The plants are propagated by seed sown in autumn and stratified through winter, by layering or root cuttings. Most plants are produced by planting seeds in deep containers and transplanting the seedlings when they are 0.6 to 1 meter (2-3 feet) tall. Suckers from horizontal roots also may be transplanted, but bare-root transplants are difficult to establish. Plants require a deep, rich soil for best production. The fleshy, edible fruit, 8 to 12 centimeters (3-5 inches) long, resembles a small potato and turns brown when ripe, usually around 150 days after flowering. The nutritious pulp is strongly aromatic, contains several large seeds, and has a luscious, custard-like consistency (Hortus, 1976; Hedrick, 1919). The fruit is usually eaten fresh but is sometimes processed (Markle et al., 1998). Sensitive individuals sometimes develop dermatitis after handling the fruit (Wyman, 1986).

banana, Hoosier *(Asimina triloba): See* **banana, custard.**

banana, Indiana *(Asimina triloba): See* **banana, custard.**

banana, Michigan *(Asimina triloba): See* **banana, custard.**

banana, Nebraska *(Asimina triloba): See* **banana, custard.**

banana, poor man's *(Asimina triloba): See* **banana, custard.**

banana tree *(Asimina triloba): See* **banana, custard.**

barberry: Berberidaceae. Barberries have been assigned the genus *Berberis,* which contains nearly 500 species, most of which are native to South America and eastern Asia. The plants are usually spiny bushes with bright yellow wood; leaves deciduous and highly colored in autumn, or evergreen, simple, and often with spiny margins; flowers yellow to red and borne singly or in clusters; fruit a red, yellow, or black berry, often exhibiting a persistent style. Most evergreen species are not hardy to the north.

Berberis vulgaris, B. canadensis, and hybrids of *B. thunbergii* and *B. vulgaris* are alternate hosts for a stem rust *(Puccinia graminis)* that also attacks wheat, oats, barley, and rye and so their planting may be restricted (Hortus, 1976).

Berberis vulgaris L. COMMON BARBERRY, JAUNDICEBERRY, PIPRAGE. This species was introduced from Europe into colonial New England, where it had become a nuisance in Massachusetts by 1754. The fruit can be preserved in sugar syrup or pickled in vinegar. Cooks in Renaissance England used the leaves to season meat (Hendrick, 1919). This species is hardy to zone 3. Deciduous species of *Berberis* tolerate drier soils but do best on rich, well drained, slightly acidic soils, in full sun or partial shade. Only maintenance pruning is required to keep the plants thrifty. Propagate the plants by seed separated from the pulp at harvest and planted in autumn to allow for winter stratification (Hartmann et al., 2002). Softwood cuttings, layering, and sucker transplant are also accepted methods of propagation. Semihardwood cuttings can be rooted under mist (Wasley, 1979). The barberry bush transplants easily (Dirr, 1998).

barberry, blue *(Mahonia aquifolium): See* **grape, Oregon.**

barberry, holly *(Mahonia aquifolium): See* **grape, Oregon.**

barberry, Mexican: *See* **grape, Oregon.**

bearberry: Ericaceae. About 50 species make up the genus *Arctostaphylos*. All are evergreen prostrate shrubs or very small trees mostly native to western North America and Central America. The common names of bearberry and manzanita are variously applied to plants in this genus. Leaves alternate, simple, leathery; urn-shaped flowers white to pink and borne in terminal racemes or panicles; fruit a smooth, red to brown-red, berry-like drupe (Hortus, 1976).

Arctostaphylos manzanita. Parry. BIGBERRY MANZANITA, COLUMBIA MANZANITA, GREENLEAF MANZANITA. Native to California and hardy to zone 7. The evergreen plants form dense clusters of white or pink flowers in early spring. Introduced in 1897.

Arctostaphylos tomentosa (Pursh) Lindl. WOOLLY MANZANITA. This plant is found along central coastal California and is hardy to zone 7. It was introduced into cultivation in 1835 (Rehder, 1947). The red berries have been used to make a cooling, refreshing, subacid drink, or dried and made into bread.

Arctostaphylos uva-ursi (L.) K. Spreng. BEAR'S GRAPE, BRAWLINS, CREASHAK, COMMON BEARBERRY, GAYUBA, HOG CRANBERRY, KINNIKINIC, KUTAI TEA, MANZANITA, MEALBERRY, MOUNTAIN BOX, SANDBERRY. This is a creeping evergreen plant. Leaves obovate to spatulate and up to 2.5 centimeters (1 inch) long, glabrous; flowers white or pink; fruit red and ripening in early summer. Plants are hardy to zone 2. It was first cultivated around 1800. The leaves were used as medicine and in the tanning industry in Sweden. The Eskimos and the Chinook, Chippewaian, and Cree Indians mixed the dried leaves with tobacco, but the dry, mealy fruit is barely edible (Hedrick, 1919). When it is used, the fruit is usually cooked, preserved, or made into jelly. The dried leaves are used as tea in Russia (Markle et al., 1998). The plants grow best in well drained acid soil in protected areas. It thrives in sandy areas where it is widely employed as a ground cover. Seed propagation is difficult due to a double dormancy. Scarify the seeds in sulfuric acid or by immersion in boiling water and subsequent soaking in cooling water for 24 hours (Emery, 1985). Select terminal semihardwood cuttings, submerge in 10 percent Clorox, dip in 8,000 parts per million (ppm) indolebutyric acid (IBA) talc, and root under intermittent mist with 21°C (70°F) bottom heat. Plants may also be propagated by layering and grafting (Hartmann et al., 2002).

bearberry, Nevada *(Arctostaphylos nevadensis): See* **bearberry.** This species is very closly realted to *Arctostaphylos uva-ursi.*

berry, baked apple *(Rubus chamaemorus): See* **blackberry.**

bignay: Euphorbiaceae. This small, slow-growing shrubby tree, *Antidesma bunius* Spreng, is native to Southeast Asia, Malaysia, and western Australia and not grown extensively in North America. Leaves glossy, somewhat leathery, alternate, oblong, with acuminate or obtuse tips and acute base, about 15 centimeters (6 inches) long and 8 centimeters (3 inches) wide; flowers small, with unpleasant fragrance, are produced from April to June near tips of the branches in terminal or axillary spikes or racemes; trees dioecious; fruit drupelike, about 1.03 centimeters (0.5 inch) in diameter and borne in racemose clusters of 20 to 40 or more. Fruit of this plant turn nearly black at maturity and are juicy, subacid, and well flavored, though ripening is uneven. Pollination may be unnecessary except for the production of viable seed. The plants are propagated by seed, cuttings, air layering, or grafting. Viable seeds may take nearly a year to germinate. The fruit has been used for jellies and wine, which some people consider especially good.

Antidesma dallachyanum Baill., *A. montanum* Blume, and *A. platyphyllum* H. Mann are similar to *A. bunius* in growth habit and fruit size and their fruit is used as those of *A. bunius* (Mowry et al., 1958).

bilberry: *See* **blueberry** and **whortleberry.**

bilberry, dwarf *(Vaccinium caespitosum): See* **blueberry.**

bilberry, Kamchatka *(Vaccinium praestans): See* **blueberry.**

bingleberry: *See* **blackberry.**

blackberry: Rosaceae. General common names include BINGLE-BERRY, BRAMBLE, BRAMBLEBERRY, CANEBERRY, CHESTERBERRY, CORYBERRY, HULLBERRY, KING'S ACEBERRY, LAVACABERRY, LAXTONBERRY, and LOWBERRY. Both blackberries and raspberries, collectively called brambles, belong to the genus *Rubus,* which contains more than 250 species. Most have prickly biennial canes with perennial roots. First-year canes are called primocanes, which become floricanes in their second and final year. Leaves alternate, mostly compound; flowers white or pink and usually clustered; fruit is an aggregate of drupelets. In blackberries and dewberries (subgenus *Eubatus*), an extremely diverse group of plants, the receptacle remains attached to the drupelets, whereas in raspberries (subgenus *Idaeobatus*) it separates from the fruit at maturity. Dewberries are simply blackberries that have a viney habit and usually require support. Both are derived from native species (Hortus, 1976).

Rubus allegheniensis T.C. Porter. ALLEGHENY BLACKBERRY, SOW-TEAT BLACKBERRY. This plant is well armed with strong, stout prickles, and many modern cultivars, such as 'Darrow', 'Eldorado', and 'Cherokee,' are derived from it. It is hardy to zone 4.

Rubus arcticus L. CRIMSON BLACKBERRY, CRIMSON BRAMBLE, CRIMSONBERRY, ARCTIC BLACKBERRY, ARCTIC BRAMBLE, NECTAR-BERRY.

Rubus baileyanus Willd. AMERICAN DEWBERRY. 'Lucretia' is derived from this species.

Rubus chamaemorus L. CLOUDBERRY, MALKA, MOLKA, SALMON-BERRY, YELLOWBERRY, BAKE APPLE, BAKED APPLE, BAKED APPLE BERRY. Hardy to zone 2.

Rubus coronarius Sweet. BRIER ROSE.

Rubus cuneifolius Pursh. SAND BLACKBERRY.

Rubus dumetorium Weihe. EUROPEAN DEWBERRY.

Rubus flagellaris Willd. AMERICAN DEWBERRY. Hardy to zone 4.

Rubus hispidus L. SWAMP DEWBERRY, RUNNING BLACKBERRY, SWAMP BLACKBERRY. Hardy to zone 4.

Rubus laciniatus Willd. CUT-LEAF BLACKBERRY, CUT-LEAVED BLACKBERRY, EVERGREEN BLACKBERRY, PARSLEY-LEAVED BLACK-BERRY. Hardy to zone 6.

Rubus macropetalus Dougl. ex Hook. BLACKBERRY, DEWBERRY.

Rubus ulmifolius Schott. EVERGREEN THORNLESS BLACKBERRY, BURBANK THORNLESS BLACKBERRY. Hardy to zone 6.

Rubus ursinus Cham. & Schlechtend. CALIFORNIA BLACKBERRY, PACIFIC DEWBERRY, PACIFIC BLACKBERRY, VEITCHBERRY. This species includes the cultivars 'Boysen' (BOYSENBERRY) and 'Young' (YOUNGBERRY). It is hardy to zone 6. L.H. Bailey places boysenberry, veitchberry, and youngberry under the separate varietas *Rubus ursinus* var. *loganbaccus* L.H. Bailey, while other authorities place youngberry under the separate species *Rubus caesius* L., also known as the EUROPEAN DEWBERRY, OLALLIEBERRY

Rubus vitifolius Cham. & Schlechtend.

U.S. production of blackberries amounted to 3,240 hectares (8,000 acres) in 1995, with 2,000 hectares (4,900 acres) of those in Oregon. Other important world blackberry production areas include Guatemala, Costa Rica, and Europe (Markle et al., 1998). In 2004 the United States produced slightly more than 5 million pounds (2,267,961 kilograms) of blackberries. Oregon led production with about 3 million pounds (1,360,777 kilograms), followed by California with about 2 million pounds (907,184 kilograms) (http://www.jan .mannlib.cornell.edu). Oregon also produced 180,000 pounds of loganberries (81,646 kilograms). Because of the extreme diversity in environmental tolerance among members of *Rubus,* blackberry cultivars can tolerate winters from –15°C (5°F) (semitrailing thornless types) to –26°C (–15°F) (a few, high-chill, erect, thorny types). 'Illini Hardy' has survived –31°C (–24°F). This makes selection of the right cultivar very important. Erect blackberries are usually thorny, though the newly released cultivars 'Navaho' and 'Arapaho' are thornless. Semitrailing thornless blackberries are heavy producers—'Chester' and 'Hull' can produce up to 22.7 kilograms (50 pounds) of fruit per plant—but have somewhat less flavor than fruit of the upright types.

Dewberries ripen earliest of all the blackberries (Poling, 1996). Trailing blackberries are hardy to –15 to –12°C (5-10°F) and have a medium chilling requirement. 'Lucretia' and 'Carolina' are two common cultivars. Erect thorny 'Darrow' and 'Illini Hardy' are hardy to –26 to –29°C (–15 to –26°F) and have a high chilling requirement; 'Arapaho', 'Cherokee', and 'Cheyenne', all hardy to –23°C (–9°F), have a medium to high chilling requirement; 'Choctaw' and 'Navaho', both hardy to –21°C (–6°F), have medium chilling requirements; 'Brazos' and 'Humble', hardy to –18 to –21°C (–1 to –6°F), have medium to low chilling requirements. Semitrailing, thornless 'Black Satin', 'Dirksen', 'Chester' (the most important cultivar of this type), and 'Hull' are all adapted to the Midwest/mid-Atlantic/mid-South. 'Georgia Thornless' is well adapted to the Gulf Coast.

Erect types of blackberries can be propagated using pencil-thick root cuttings dug and planted in early spring, and by transplanting suckers. Trailing and semitrailing types are propagated by tip layers.

Choose sites in full sun and with good air and water drainage, especially for trailing types and other early ripening cultivars. Use wind breaks in open, windy areas to minimize cane dessication. Do not plant blackberries in ground recently cropped to potatoes, tomatoes, peppers, or eggplants due to potential for infection from soil-borne wilt pathogens, or where peach, apple, grape, or other bramble plantings recently stood because of the potential for crown gall infection. Plant blackberries away from wild stands and adjust soil pH to slightly acid prior to planting. Apply approximately 0.5 kilogram (1 pound) of a 10 percent nitrogen fertilizer per 7 meters (20 feet) of row in early spring to stimulate rapid growth. Beginning in the second year and continuing yearly, sidedress after harvest with about 1 kilogram (2 pounds) of ammonium nitrate per 30 meters (100 feet) of row to provide for vigorous growth. Semitrailing types may need slightly more fertilizer. Do not apply nitrogen fertilizer later than early July in any year (Moore and Skirvin, 1990; Tompkins, 1977). Plant erect and semitrailing types about 3 meters (10 feet) apart and trailing types about 2 meters (6 feet) apart. Blackberries produce new shoots from their crowns. The "primocanes" produce laterals the first year. In the second year, the laterals form small shoots that produce fruit near their tips. Second-year canes are called "floricanes."

Proper pruning is very important for good production. Remove the floricanes as soon as fruiting is complete since these canes will fruit only once, then die. Erect types may grow to heights of 5 meters (17 feet). Allow suckers to develop in a row about 20 centimeters (8 inches) wide and pull out those that arise outside this row. Top primocanes in midsummer at 75 centimeters (2.5 feet) height to force strong lateral growth. Cut back laterals to 30 to 40 centimeters (12-16 inches) in length during late winter pruning and thin healthy canes to stand about 6 centimeters (2.5 inches) apart in the row. For semitrailing and trailing thornless types pruning is similar to that for erect types, except that the trailing canes must be trellised as soon as they reach about 1.5 meters (5 feet) in length. Keep the crop weed-free by mulching or with frequent hoeing. Good site sanitation will help keep the planting pest free (Gough, 1997). Harvest usually begins about 60 to 90 days after bloom, and the crop may be picked at least twice weekly during the peak of the season. Pick the fruit in the morning when fully ripe (color appears dull) and handle carefully to avoid crushing. Blackberries stored at 1°C (34°F) and 90 to 95 percent relative humidity should remain in good condition for only two to three days (Lutz and Hardenburg, 1968). Blackberries are used fresh, frozen, canned, and in desserts, salads, preserves, jams, wines, and brandy. The fresh berries contain 85 percent water and 10 percent carbohydrates and are high in minerals, vitamin B, vitamin A, and calcium (Poling, 1996).

blackcap *(Rubus occidentalis): See* **raspberry.**

blackthorn *(Prunis spinosa): See **Prunus.***

blaeberry: Name variously applied to species of *Vaccinium,* particularly *V. myrtillus. See* **blueberry.**

blueberry: Ericaceae. The genus *Vaccinium* contains about 150 species, the major ones of which are listed below. Most are deciduous or evergreen shrubs native to the northern hemisphere; leaves simple, alternate, bright autumn color; flowers solitary or in clusters, urn-shaped, white to pink; fruit a many-seeded berry. The fruit of nearly all species is edible and has been used for centuries around the world (Hortus, 1976; Hedrick, 1919).

Vaccinium angustifolium Ait. LATE SWEET BLUEBERRY, LOWBUSH BLUEBERRY, LOW SWEET BLUEBERRY, SWEET-HURTS. Plants reach a height of up to 30 centimeters (12 inches) and are hardy to zone 2. This species is usually not cultivated but rather is harvested from wild stands primarily in Maine, which produced nearly the entire U.S. crop (33,000 tons, or 29,937 metric tons, worth $21 million) in 1995 (Markle et al., 1998; Davis, 1996). In 2004, Maine produced about 23,000 tons (20,965 metric tons) of wild blueberries (http://www .jan.mannlib.cornell.edu). Canada harvested fruit from 5,790 wild hectares (14,295 acres) in 1995. Fruit is harvested approximately two to four months after bloom and all but about 1 percent is used in the processing trade.

Vaccinium ashei Reade. RABBIT-EYE BLUEBERRY. This species produces a suckering shrub up to 1 to 6 meters (3-20 feet) in height.

Vaccinium atrococcum (A. Gray) A. Heller. BLACK HIGHBUSH BLUEBERRY. Shrubs of this species grow up to 3 meters (10 feet) high and are hardy to zone 4.

Vaccinium caesium Greene. DEERBERRY, SQUAW HUCKLEBERRY. This is a shrub up to 1 meter (3 feet) high.

Vaccinium caespitosum Michx. DWARF BILBERRY. A shrub to 30 centimeters (12 inches) high and hardy to zone 2.

Vaccinium corymbosum L. HIGHBUSH BLUEBERRY, SWAMP BLUE-BERRY, WHORTLEBERRY. This shrub grows to 4 meters (14 feet) high and is hardy to zone 4.

Vaccinium crassifolium Andr. CREEPING BLUEBERRY. An evergreen, prostrate shrub.

Vaccinium deliciosum Piper. Shrub to 30 centimeters (12 inches) high.

Vaccinium elliottii Chapm. ELLIOTT'S BLUEBERRY. Shrub to 3 meters (10 feet) high.

Vaccinium membranaceum Dougl. BLUE HUCKLEBERRY, MONTA-NA BLUE HUCKLEBERRY, MOUNTAIN BLUEBERRY, THIN-LEAF HUCK-LEBERRY. This shrub up to 1.5 meters (5 feet) high is hardy to zone 6.

Vaccinium mortinia Hook. F. MORTINIA. An evergreen procumbent shrub to 1 meter (3 feet) high.

Vaccinium myrsinites Lam. An evergreen shrub to 0.6 meter (2 feet) and hardy to zone 7.

Vaccinium myrtilloides Michx. SOURTOP, SOURTOP BLUEBERRY, VELVET-LEAF BLUEBERRY. A shrub to 30 centimeters (12 inches) high, hardy to zone 2. The fruit and plants of this species are similar to those of the lowbush blueberry though the plants are slightly more shade tolerant. The fruit is glabrous, blue-black, sweet, and approximately 8 millimeters (0.3 inch) in diameter.

Vaccinium myrtillus L. BILBERRY, BLAEBERRY, DWARF BILBERRY, SOURTOP, VELVET-LEAF BLUEBERRY, WHINBERRY, WHORTLEBERRY. This is a shrub to 60 centimeters (24 inches) in height that is hardy to zone 4. This species is similar to *V. myrtilloides* and both are popular in Scotland, where the fruit is used in wine and preserves. It is also used medicinally to treat cystitis (Markel et al., 1998).

Vaccinium occidentalis A. Gray. WESTERN BLUEBERRY. A shrub to 1 meter (3 feet) high.

Vaccinium ovalifolium Sm. Mathers. A shrub to 5 meters (17 feet) high, hardy to zone 5.

Vaccinium ovatum Pursh. CALIFORNIA HUCKLEBERRY, EVER-GREEN HUCKLEBERRY, SHOT HUCKLEBERRY. An evergreen shrub to 3 meters (10 feet) high and hardy to zone 7. This is a major source of the "huckleberry" or "leatherleaf" used in the florist trade.

Vaccinium praestans Lamb. KAMCHATKA BILBERRY. A deciduous shrub to 15 centimeters (6 inches) in height, hardy to zone 4.

Vaccinium scoparium Leib. GROUSEBERRY, LITTLE-LEAF HUCK-LEBERRY. A diminutive deciduous shrub to 40 centimeters (16 inches).

Vaccinium stamineum L. DEERBERRY, SQUAW HUCKLEBERRY. This is a deciduous shrub to 3 meters (10 feet) in height.

Vaccinium uliginosum L. BOG BILBERRY, MOORBERRY. A shrub to 60 centimeters (24 inches) that is hardy to zone 2.

Vaccinium vacillans Torr. EARLYSWEET BLUEBERRY, LOW BIL-BERRY, LOW BLUEBERRY, LOW SWEET BLUEBERRY, SUGAR HUCKLE-BERRY. This is a deciduous shrub to 1 meter (3 feet) in height.

Commercial production figures for blueberries often combine the *V. ashei* and the *V. corymbosum* crops. In 2004 the United States produced about 229,000,000 pounds (103,872,652 kilograms), with highbush production mainly centered in New Jersey (39,000,000 pounds or 917,690,102 kilograms), Michigan (80,000,000 pounds or 36,287,389 kilograms), North Carolina, Oregon, and Washington and rabbiteye production in North Carolina, Georgia, Florida, and Arkansas (www.jan.mannlib.cornell.edu). Other areas of the world producing significant amounts of these species include Canada, with 79,161 metric tons (87,262 tons) produced in 2004, and Poland, with 16,500 metric tons (18,188 tons) (http://www.faostat.fao.org). Three species are commonly grown commercially under the name of "blueberry"—*V. angustifolium, V. ashei,* and *V. corymbosum.*

V. angustifolium, the lowbush blueberry, is semidomesticated by periodic pruning and fertilizing of wild stands, although some improved cultivars have been developed. Most of the fruit is used in the canning and culinary trade as pastry filling, etc.

V. ashei, the rabbiteye blueberry, is a highbush type grown in the southern United States.

Most blueberries sold as fresh fruit belong to the highbush blueberry, *V. corymbosum.* This species can be subdivided into southern highbush and northern highbush cultivars. The "halfhighs" are a subgroup of the northern highbush type and generally have some lowbush blueberry parentage. They reach a height of about 1 meter (3 feet) and are particularly well adapted to the northern boundaries of good blueberry country.

Blueberries form their flower buds at the tips of the current year's wood in mid- to late summer of the year preceding bloom. Berries are borne in clusters of five to twelve fruit. Highbush blueberry and rabbiteye (R) cultivars are adapted to various regions of the country as follows:

> North Florida and Gulf Coast: 'Avonblue', 'Flordablue', 'Climax' (R), 'Bluebelle' (R).
> Coastal plain of Georgia, South Carolina: 'Flordablue', 'Southland', 'Climax' (R), 'Woodard' (R).
> Upper Piedmont: 'Bluetta', 'Patriot', 'Bluebelle' (R), 'Southland' (R).

Southern Virginia, Tennessee, lower Ohio Valley, lower Southwest: 'Croatan', 'Bluecrop', 'Woodard' (R), 'Climax' (R).

Mid-Atlantic, Ozark highlands, Midwest: 'Bluetta', 'Blueray', 'Bluecrop'.

New England and Great Lakes: 'Earliblue', 'Patriot', 'Coville'.

Halfhigh cultivars include 'Northblue', 'Northcountry', and 'Northsky'. The lowbush blueberry is usually harvested from wild stands but some improved cultivars have been released, including 'Augusta', 'Blomidon', 'Chignecto', and 'Fundy' (Carter and St. Pierre, 1996). Blueberries are generally considered self-fruitful, though some cases of self-unfruitfulness exist. Therefore, always set out a mixed planting of several cultivars. The plants are pollinated by bees, particularly bumblebees.

Both lowbush and highbush blueberry types are propagated by pencil-thick hardwood, semihardwood, and softwood cuttings taken from the middle portions of sound wood where no flower buds are present (Moore and Ink, 1964). The latter two types of cuttings may be dipped in 4.5 percent IBA talc formulation to stimulate rooting (Gough, 1994). They also may be propagated by mound layering, leaf-bud cuttings, clump division, and budding (Ballington et al., 1989). Lowbush blueberries are allowed to spread naturally in wild stands by rhizomes. Rabbiteye types are generally propagated by softwood cuttings and sucker division.

All blueberries require full sun and moist, acid soil (pH 5) high in organic matter. The delicate hairlike roots do not penetrate heavy compacted soils. Avoid wet, poorly drained soils and sites with poor air drainage. Wet soils can harbor the *Phytophthora* root rot fungus (Gough, 1996). Further, iron-deficiency-induced chlorosis is often a problem when soil pH rises above six. Symptoms include stunting, interveinal yellowing (chlorosis) of the leaves, and poor plant growth. Foliar applications of iron chelate or ferrous sulfate at a rate of 28 and 114 grams (1-3.5 ounces) per plant offer temporary correction. Permanent correction involves lowering soil pH through the use of acidifying agents such as sulfur and iron sulfate.

Plant two- to three-year-old plants in early spring. Plant spacing varies according to the size of the plantation and the species. Lowbush blueberries are usually established in solid plantings, while halfhighs are spaced about 1 meter (3 feet) apart in rows spaced 2 meters (6 feet) apart. General spacing of highbush plants is 1.5 to 2 meters (4-6 feet) apart in rows spaced 2 to 3 meters (6-9 feet) apart. Rabbiteye plants often are spaced a bit farther apart to accommodate their larger size.

In general, blueberries have a lower requirement for nutrients than many other fruit crops. Avoid fertilizers containing the chloride ion, to which blueberry is very sensitive. About a month after planting sidedress highbush blueberries with 1 ounce (31 grams) of a 10 percent nitrogen fertilizer. Increase fertilizer amounts each year so that mature plants (after six years in the field) receive about 0.5 kilograms (1 pound) per plant, half applied in early spring and the rest about six weeks later. Rabbiteye blueberries are very sensitive to overfertilization; do not fertilize them the first year. All blueberries do well with organic fertilizer sources, such as well rotted barnyard manure or cottonseed meal (Gough, 1994; Gough, 1996). Organic mulches have proven highly successful for blueberries. Sawdust, peat, pine needles, and bark all reduce soil compaction, winter heaving, and root damage while at the same time increasing soil moisture and the survival of new plants. Straw mulch tends to attract rodents with subsequent vole damage to the bushes occurring in winter.

Because blueberry forms flower buds only on new wood, proper pruning is important to keep the bush vigorous and to remove older, unproductive wood. Prune highbush plants annually in the early spring by removing canes older than six years of age and canes that arise outside the crown area of approximately 30 to 40 centimeters (12-16 inches). Allow no more than about eight to ten healthy canes per bush. Rabbiteye bushes may need pruning only once every few years. Lowbush blueberries are usually burned or mown to a 2.5 centimeter (1 inch) height every two to three years as an effective means of pruning. Vigorous new growth arises from the rhizome after the tops of the plants have been removed.

Blueberry plants reach maturity and full production after about six to eight years in the field, though they may begin to bear in their second year. Ripening is not concentrated so individual fruit is harvested when ripe and harvesting occurs over a few weeks, beginning two to four months after bloom. A blueberry fruit is ripe about five to seven days after it turns blue. During this time it may also increase in size by up to a third. Therefore, delay harvest accordingly so that the fruit will reach their greatest size (Gough, 1994; Gough, 1996). Lowbush blueberries are harvested by raking the fruit from the plants by hand or with mechanical harvesters. Highbush and rabbiteye are harvested by hand or mechanically.

Blueberries in good condition may be stored for a week or two at 1°C (34°F) and 90 to 95 percent relative humidity (Lutz and Hardenburg, 1968). One cup of fresh berries contains about 21 grams (0.7 ounce) of carbohydrate, 1 gram (0.03 ounce) of protein, and 0.5 gram (0.0006 ounce) of fat, 19 milligrams of vitamin C, 145 international units of vitamin A, and only 85 calories. The fruit also contains measurable amounts of ellagic acid, known for its cancer-fighting qualities (Hortus, 1976). Conner and colleagues (2002) reported that not only genotype but environment alters the content of beneficial compounds, such as anthocyanins and antioxidants, in blueberry fruit.

blueberry, Rocky Mountain *(Amelanchier alnifolia): See* **serviceberry.**

Blue-tangle *(Gaylussacia frondosa): See* **huckleberry.**

bohemian berry *(Fragaria moschata): See* **strawberry.**

box, mountain *(Arctostaphylos uva-ursi): See* **bearberry.**

boxberry *(Gaultheria procumbens): See* **wintergreen.**

boxthorn: *See* **wolfberry.**

boysenberry: A cultivar of *Rubus ursinus* or a hybrid of loganberry × dewberry. Discovered by Ralf Boysen in California in 1920. It is larger, more tart, and earlier than loganberry or youngberry (Darrow, 1955). *See* **blackberry.**

bramble: *See* **blackberry.**

brambleberry: *See* **blackberry** and **raspberry.**

brawlins *(Arctostaphylos uva-ursi): See* **bearberry.**

breadfruit, Mexican *(Monstera deliciosa): See* **monstera.**

brier, rose: *See* **rosehips.**

buckthorn, sea: Elaeagnaceae. One of two species in the genus *Hippophae*. Both are dioecious spiny shrubs or small trees native to Europe and Asia; leaves alternate, willow-like, deciduous; flowers inconspicuous, yellow; fruit edible.

Hippophae rhamnoides L. COMMON SEA BUCKTHORN, SALLOW THORN. The plant grows to a height of 10 meters (33 feet) and is hardy to zone 3. The bright orange fruit is quite acid, but edible and highly valued in China and Russia. Any soil is acceptable, but plants do better in sandy, infertile soil with moist subsoil. Plants should be in full sun for proper fruiting. Be sure to have both male and female plants close by. A ratio of one male to six female is acceptable. The species is propagated by autumn-sown seed, hardwood cuttings (difficult), root cuttings, layering, and sucker transplant. Plants are difficult to transplant (Hortus, 1976; Wyman, 1986; Wyman, 1969; Dirr, 1998). The fruit is an excellent source of vitamin C.

buffaloberry: Elaeagnaceae. There are three species of dioecious shrubs in the genus *Shepherdia* commonly called by this name, only one of which is grown for its fruit. The genus name commemorates the English botanist John Shepherd. *Shepherdia* is closely related to the genus *Elaeagnus* and so is related to Russian and autumn olive. Plants have brown or silvery scales and are native to North America. Leaves opposite, simple; flowers yellow; fruit bright red-orange, drupelike, 3 to 5 millimeters (0.1-0.2 inch) in diameter. All members of the genus *Shepherdia* are dioecious, making cross-pollination necessary. They are adapted to dry, rocky, or stony soils and dry, alkaline conditions (Hortus, 1976) and were first cultivated in 1818. The plant was introduced into commercial cultivation in Big Horn City, Wyoming, in the fall of 1890 (Remlinger and St. Pierre, 1995).

Shepherdia argentea (Pursh) Nutt. BEEF-SUET TREE, BUFFALO-BERRY, CRUCIFIXION BERRY, NEBRASKA CURRANT, RABBITBERRY, SILVERBERRY, SILVER BUFFALOBERRY, THORNY BUFFALOBERRY, WILD OLEASTER. This shrub grows to 6 meters (20 feet) high and is hardy to zone 2 (Wyman, 1969). Native Americans and early settlers made a conserve from the fruit for use with buffalo meat, giving the plant its name. A paste or pudding of buffaloberry fruit along with flour made of prairie turnip *(Psoralea esculenta)* often was served with the meat as well. The fruit was used in pemmican, soup, and for juice.

Flowers are borne singly or in clusters in the axils of two-year-old branches. Pollination is usually accomplished by bees and perhaps small flies. Two cultivars are available: 'Gold-eye', a yellow-fruited cultivar developed in Manitoba, and 'Sakakawea', developed by the U.S. Department of Agriculture. The cultivar 'Xanthocarpa' belongs to the species *S. canadensis* (Remlinger and St. Pierre, 1995).

Plants are usually propagated by autumn-sown seed. Suckers are easily transplanted, and cuttings of the yellow-fruited genotypes, collected in late July, are fairly easy to root after treatment with 0.8 percent IBA/talc and thiram and striking in a 1:1 sand:perlite mixture under mist (Remlinger and St. Pierre, 1995). The plants do best on medium-textured, well drained, moist soils but survive under drier conditions. They tolerate drought, saline soils and intermittent flooding, but not shade. One male plant to every six female plants will provide adequate cross-pollination. Male plants have larger, stouter winter buds and their blossoms are about twice the size of female blossoms. Plant two-year-old plants slightly deeper than they were in the nursery and about 4 meters apart in rows spaced 5 meters (16 feet) apart if you wish to maintain individual plants.

Fertilizer recommendations have not been worked out for this species. It does fix atmospheric nitrogen, suggesting the need for minimal applications of that nutrient.

The plants begin to bear fruit four to six years after planting and the fruit begins to ripen a bit over 100 days after bloom, ripening in the order in which the flowers were pollinated. It is considered ripe when fully yellow or red. The astringent fruit is often harvested after a frost when the astringency abates and sugar content increases. Harvesting is made difficult by the presence of thorns, the uneven ripening, and the tendency for the fruit to adhere tightly to the bush. The use of heavy gloves remedies the thorny problem but makes handling the small berries difficult. Traditionally, a cloth is laid on the ground beneath the bush and the fruit is beaten off the branches when the temperature is about −10°C (14°F), since the frozen fruit is more easily detached (Remlinger and St. Pierre, 1995).

Process the fruit immediately after harvest; overripe fruit attain a strong flavor and unpleasant odor and become sticky. Consume raw berries only in moderation because they contain saponins, the agent that causes their bitter flavor and also can cause severe gastrointestinal upset when consumed in large quantities (Remlinger and St. Pierre, 1995). Buffaloberry fruit is high in vitamin C (150 milligrams/100 grams of fruit). A relative, *S. canadensis,* is known to accumulate mercury. Buffaloberry might do the same thing and so could become a health problem in contaminated areas.

bullace *(Vitis rotundifolia): See* **grape.**

cactus, spineless *(Opuntia ficus-indica): See* **pear, prickly.**

calabash, sweet *(Passiflora maliformis): See* **passionfruit.**

caneberry: *See* **blackberry.**

capiton berry *(Fragaria moschata): See* **strawberry.**

capron: *(Fragaria moschata): See* **strawberry.**

carissa: Apocynaceae. CARISSA, NATAL-PLUM, AMATUNGULA. This is a dense tropical shrub, reaching heights of up to 4.5 meters (15 feet), related to **karanda,** though native to South Africa. Leaves dark green, leathery; branches heavily armed with two-pointed spines; flowers fragrant, white, solitary, about 5 centimeters (2 inches) wide and highly conspicuous; fruit dark red, ovoid, 2.5 to 5 centimeters (1-2 inches) long; pulp reddish, numerous small seeds. The plant is widely adapted to Florida soils and is used extensively in the southern part of the state as an ornamental shrub. It is usually propagated by seeds, but seedlings are slow-growing. Layering, marcottage, and cuttings are also successful methods of rooting this species. The use of bottom heat in the cutting flats may speed root formation.

cas: *See* **guava.**

cassis *(Ribes nigrum):* See **Ribes.**

ceriman *(Monstera deliciosa):* See **monstera.**

cerza: *See* **cherry, Barbados.**

checkerberry *(Gaultheria procumbens):* See **wintergreen.**

checkertree *(Sorbus torminalis):* See **chequers.**

chequers: Rosaceae. About 85 species, many of them shrubs of the northern hemisphere, comprise the genus *Sorbus.* Leaves alternate, simple or pinnate; flowers white, in terminal clusters, bisexual; fruit a small pome.

 Sorbus torminalis (L.) Crantz. CHECKERTREE, WILD SERVICE TREE. Hardy to zone 6 but does poorly further south. The fruit is a small, brown pome about 2 centimeters (1 inch) in length. This species is usually propagated by seed sown in the autumn. Cuttings are difficult to root, but long, softwood cuttings of related species taken from forced stock plants root well at the junction of current and one-year-old wood (Hanson, 1990). Transplant balled and burlapped plants into slightly acid soils. Compacted soils, high soil pH, high summer temperatures, and low or fluctuating winter temperatures predispose this species in particular, and the genus in general, to borers and cankers. Excess nitrogen fertilizer may predispose the genus to fireblight (Hortus, 1976; Dirr, 1990). Practice maintenance pruning only. Harvest the fruit after a frost or when it is overripe, as medlars are harvested. The fruit is made into wine or jellies or dried. The bark has been used to tan leather (Markle et al., 1998).

chereese: *See* **cherry, Barbados.**

cherry *(Prunus incisa):* See **Prunus.**

cherry, Barbados: Malpighiaceae. ACEROLA, CERZA, CHEREESE, FRENCH CHERRY, GARDEN CHERRY, JAMAICA CHERRY, NATIVE CHERRY, PUERTO RICAN CHERRY, WEST INDIAN CHERRY, WEST IN- DIES CHERRY. The entry common name is also applied to members of the genus *Eugenia.* About 30 species of evergreen shrubs and trees, mostly native to tropical America, make up the genus *Malpighia*. Leaves opposite, simple; flowers white or red, sessile or cymose; fruit a red, orange, or purple drupe borne singly or in axillary clusters of two or three with three triangular seeds (Hortus, 1976).

Malpighia glabra Millsp. (*M. punicifolia* L.). BARBADOS CHERRY. An evergreen shrub or small tree 3 to 5 meters (10-17 feet) and sometimes 6 meters (6.5 feet) in height grown for its edible, waxy, cherry-sized fruit. The fruit has a single large seed, smooth, crimson skin, and three prominent ribs, and contains the highest vitamin C content of any common fruit, having a concentration of the vitamin 60 times greater than that of orange juice. Only the hips from *Rosa rugosa* contain a higher concentration of this vitamin. There has been some taxonomic confusion regarding this species. Though we now know that Barbados cherry may actually be a hybrid of *M. glabra* × *M. punicifolia,* the more commonly accepted botanical name of *M. punicifolia* L. has generally replaced *M. glabra* Millsp. The plant was introduced into the United States in the late nineteenth century, where it was occasionally planted in south or central Florida. During World War II the trees were planted in schoolyards in that area to provide a source of vitamin C for schoolchildren. The plant was quickly exploited and marketed for its very high content of vitamin C under its Puerto Rican name "acerola." By 1954, Hawaii had an 800 hectare (2,000 acre) plantation and there were some 30,000 trees in commercial groves on Puerto Rico. Interest in the fruit subsided when it could not compete with synthetically produced vitamin C. The raspberry-flavored fruit can be eaten fresh or processed into jams and preserves. The plant is hardy only to zone 10 (Hedrick, 1919; Wyman, 1986).

Puerto Rico had about 200 hectares (500 acres) in production in 1980. Hawaii, other Caribbean countries, and northern South America also produce fruit for the jelly and vitamin trade (Morton, 1987).

A few cultivars have been introduced. 'Florida Sweet' was released by the University of Florida in 1956. The Puerto Rican clone 'B-15' is a high yielding selection. Bees are the principle pollinators. In Hawaii, pollination by the natural bee population is often inadequate and application of 100 ppm IBA has been used experimentally to induce higher fruit set.

The plant is propagated by seeding, cuttings, layers, or grafting. Germination percentages are low, ranging from 5 to 50 percent. Air layering and side-veneer, cleft, or modified crown grafting are useful, but cuttings are the most popular means of propagation. Pencil-thick cuttings with two to three leaves attached and 20 to 25 centimeters (8-10 inches) long are dipped in rooting hormone and struck in sand under constant mist for 60 days. Rooted cuttings are transferred to the nursery bed and grown under shade for a year before field setting.

The plant must have well drained soils with pH above 5.5, a pH of 6.5 being nearly ideal. Though the species is grown in hot, tropical areas with medium to high rainfall, mature trees can survive very brief exposure to –2°C (28°F), while young plants are killed at –1°C (30°F).

The plants need minimal care. In Puerto Rico, plants are given two, 0.5 to 1 kilogram (1-2 pounds) applications of a complete fertilizer (8-8-15) annually for the first several years. The amount is increased so that older trees receive 1 to 2 kilograms (3-5 pounds) annually. In Florida, 0.4 kilogram (1 pound) of 10-10-10 is applied in February for each year of plant growth, and another 0.4 kilogram (1 pound) of 4-7-5-3 for each year of age up to 10 years is provided in May, July, and September. Thereafter, 2 kilograms (5 pounds) of 6-4-6-3 per plant is given in late winter and 3.5 kilograms (10 pounds) per tree in each of the summer feedings (http://www.hort.purdue.edu/newcrop/morton/barbados_cherry.html).The plants are thinned by judicious pruning to maintain fruit size and may require applications of some minor nutrients to correct deficiencies.

The first crop is produced in the third or fourth year and the plants will continue bearing well for about 15 years. Fruit is ripe about three weeks after flowering, though the harvest season varies according to the weather. In Florida, the Bahamas, Puerto Rico, and Hawaii, bushes may produce a spring crop in May and some small crops throughout the year until December. In the absence of spring rains, only a heavy December crop will be produced. The fruit ripens only in December and January in some parts of Africa. Individual mature plants yield from 11 to 30 kilograms (30-80 pounds) of fruit, though yields vary considerably by location.

Fruit should be stored no more than three days at 7.22°C (45°F). Since refrigerated juice loses 20 percent of its ascorbic acid within 18 days it should be kept no longer than a week. The fruit is eaten out of hand or stewed with sugar for dessert. The juice reduces the speed of oxidation and darkening of fruit in salads and can be used for punch or sherbet. It makes excellent jelly, jam, and other preserves, but upon cooking and slow oxidation the bright red color turns a brown-red. Wine made from the fruit retains 60 percent of its ascorbic acid. Medium-ripe fruit contains 3,300 milligrams of ascorbic acid per 100 grams of fruit, while the unripe fruit may contain up to 4,676 milligrams/100 grams. The fruit is considered useful to people with liver ailments, diarrhea and dysentery, colds, and coughs. However, ingestion of excessive amounts may lead to intestinal inflammation and some people develop dermatitis from the minute stinging hairs on the leaves and petioles.

cherry, Barbados ground *(Physalis peruviana): See* **cherry, ground.**

cherry, bladder: *See* **cherry, ground.**

cherry, Brazil *(Eugenia uniflora): See* ***Eugenia.***

cherry, bush: *See* ***Prunus.***

cherry, cayenne *(Eugenia uniflora): See* ***Eugenia.***

cherry, European ground *(Prunus fruticosa): See* ***Prunus.***

cherry, French: *See* **cherry, Barbados.**

cherry, garden: *See* **cherry, Barbados.**

cherry, ground *(Prunus fruticosa): See* ***Prunus.***

cherry, ground: Solanaceae. BLADDER CHERRY, CHERRY TOMATO, CHINESE LANTERN PLANT, GOLDENBERRY, HUSK TOMATO, JAMBERRY, MILTOMATE, PERUVIAN GROUND CHERRY, PONA, STRAWBERRY TOMATO, TOMATILLO GROUND CHERRY, WEST INDIAN CHERRY. About 80 species make up the genus *Physalis,* a few of which are grown in home gardens for their edible fruit which is made into preserves and pickles. Some species are grown in limited commercial quantities in New Zealand (Logan, 1996). Most are upright or trailing herbs from the New World, annual in the north and perennial in the tropics. All species grow best in warm, sunny areas with rich loam soils, with culture similar to that of tomato. Fruit is enclosed in a papery husk, vary from red to yellow when mature, and may reach 2.5 centimeters (1 inch) in diameter. From seed to harvest usually takes three to four months, with bloom to maturity taking two to three months. Fruit ripens unevenly and harvest may last up to two months.

Physalis alkekengi L. ALKEKENGI, JAPANESE LANTERN, STRAW-BERRY GROUND CHERRY, STRAWBERRY TOMATO, WINTER CHERRY. This perennial plant with red fruit is sometimes cultivated as an annual.

Physalis ixocarpa Brot. HUSK TOMATO, JAMBERRY, MEXICAN HUSK TOMATO, MILTOMATE, TOMATILLO, TOMATILLO GROUND CHERRY. The fruit, sometimes reaching 7 centimeters (2.8 inches) in diameter, is purple, smooth-skinned, and somewhat sticky. It is used to flavor meat and chili in Mexico and Latin America and there is limited commercial production in California.

Physalis peruviana L. BARBADOS GROUND CHERRY, CAPE GOOSE-BERRY, CHERRY TOMATO, GOLDENBERRY, GROUND CHERRY, GOOSE-BERRY TOMATO, PERUVIAN CHERRY, PERUVIAN GROUND CHERRY, POHA, STRAWBERRY TOMATO, WINTER CHERRY. The small, yellow fruit is grown in Washington and south Florida for local consumption (Schreiber, 1995; Stephens, 1988). About 10 acres (4 hectares) of the fruit is grown in Hawaii.

Physalis pruinosa L. DWARF CAPE GOOSEBERRY, STRAWBERRY TOMATO. The fruit is yellow and eaten raw or cooked.

Physalis pubescens L. DOWNY GROUND CHERRY, GROUND CHERRY, STRAWBERRY TOMATO. This species is native to North America and produces small, mild-flavored, yellow fruit.

cherry, Hansen's bush *(Prunus besseyi): See **Prunus**.*

cherry, Indian *(Ziziphus mauritiana): See* **jujube.**

cherry, Jamaica: *See* **cherry, Barbados.**

cherry, Mongolian *(Prunus tomentosa): See **Prunus**.*

cherry, Nanking (Manchu) *(Prunus tomentosa): See **Prunus**.*

cherry, native: *See* **cherry, Barbados.**

cherry-of-the-Rio-Grande *(Eugenia aggregata):* See *Eugenia.*

cherry, Peruvian *(Physalis peruviana):* See *Eugenia. Also see* **cherry, ground.**

cherry, Peruvian ground *(Physalis peruviana):* See **cherry, ground.**

cherry, pitanga: See *Eugenia.*

cherry, Puerto Rican: See **cherry, Barbados.**

cherry, Surinam *(Eugenia uniflora):* See *Eugenia.*

cherry, West Indian: See **cherry, ground.**

cherry, winter *(Physalis alkekengi):* See **cherry, ground.**

chesterberry: See **blackberry.**

Chinese gooseberry *(Actinidia chinensis):* See **kiwifruit.**

chocolate vine *(Akebia quinata):* See *Akebia.*

chokeberry: Rosaceae. The genus *Aronia* (sometimes considered a subgenus of *Pyrus*) contains a few species of low deciduous shrubs up to 3 meters (10 feet), native to North America. Leaves alternate, simple; flowers small, pink or white in terminal clusters; fruit a small berrylike pome. The name "chokeberry" is taken from the astringent taste of the fruit.

Aronia arbutifolia (L.). CHOKEBERRY, RED CHOKEBERRY. Plants grow up to 3 meters (10 feet) high and become leggy with age, producing abundantly bright red berries less than 1 centimeter (0.5 inch) in diameter. The plant was introduced into cultivation about 1700 and is hardy to zone 5.

Aronia melanocarpa (Michx.) Elliott. BLACK CHOKEBERRY. Bushes up to 1 meter (3 feet) high. The plant was first cultivated about 1700 and is hardy to zone 3. This species is native to northeastern North America where it is gathered from the wild. There is some production in Russia (Facciola, 1990). The astringent fruit high in pectin is shiny black or red, up to 6 millimeters (0.25 inch) in diameter, and borne in large clusters. It is used to color food and may be eaten out of hand or processed into stews, jelly, and beverages. One cultivar, 'Nero', has been selected.

Aronia prunifolia (Marsh.) Rehd. PURPLE-FRUIT CHOKEBERRY. Plants grow up to 4 meters (14 feet) in height, but otherwise resemble those of *A. arbutifolia*. This species was introduced into cultivation about 1800 and is hardy to zone 5 (Hortus, 1976; Rehder, 1947).

Aronia grows well on a wide range of soils and is propagated by clump division, softwood cuttings, or autumn-planted seed (Wyman, 1986). Softwood cuttings root best when the basal cut is made 1 centimeter (0.5 inch) below a node. In lieu of autumn planting, stratify seed for 90 days at 2°C (35°F) before spring planting (Dirr, 1998).

chokecherry *(Prunus virginiana): See* **Prunus***.*

chuckley-plum *(Prunus virginiana): See* **Prunus***.*

cloudberry *(Rubus chamaemorus): See* **blackberry***.*

coryberry: *See* **blackberry***.*

cowberry: *See* **cranberry, highbush***.*

crakeberry *(Empetrum nigrum): See* **crowberry***.*

crampbark *(Viburnum opulus): See* **cranberry, highbush***.*

cranberry: Ericaceae. Both American and European cranberries belong to the genus *Vaccinium,* making them closely related to the blueberry and lingonberry.

Vaccinium macrocarpon Ait. AMERICAN CRANBERRY, CRANBERRY, LARGE CRANBERRY. This is an evergreen, mat-forming plant with leaves oblong-elliptic, flowers pink and borne in lateral clusters. The fruit is red, round, or elliptical, and about 2 centimeters (1 inch) diameter. The plants are found in acid bogs and swamps as far north as zone 2.

Vaccinium oxycoccos L. EUROPEAN CRANBERRY, SMALL CRAN-BERRY. This plant forms an evergreen, creeping mat and produces a red berry about 1 centimeter (0.5 inch) diameter. It is hardy to zone 2. In 2004 the United States produced 298,000 metric tons (328,488 tons) of cranberries (http://www.faostat.fao.org). Massachusetts produced 1,804,000 hundredweight barrels (81,828,063 kilograms), led only by Wisconsin, with 3,480,000 barrels (157,850,144 kilograms). Maine, Michigan, Delaware, New Hampshire, Minnesota, Rhode Island, and New York reported minimal acreage. Other world production areas include Canada, producing over 62,194 barrels (2,821,072 kilograms) in 2004, and Belarus, producing 23,000 barrels (1,043,262 kilograms). The fruit is borne singly on upright shoots 15 to 20 centimeters (6-8 inches) long. Cultivars include 'Howes', 'Early Black', and 'McFarlin'.

The cranberry of commerce and the only cultivated species is *Vaccinium macrocarpon.* Its cultivation is highly specialized, expensive, and restricted to acid (pH 3.2-4.5) bogs in the northern United States and Canada. Native vegetation on sites suitable for cranberry production include leatherleaf (*Chaemadaphne calyculata* Moench), sphagnum moss (*Sphagnum* spp.), and bog laurel (*Kalmia* spp.) (Dana, 1990). Bogs are flooded for irrigation, pest control, as protection against winter damage, and during harvest operations. Flooding for frost and winter protection has been superceded in many bogs by more efficient sprinkler systems. For best plant growth the water table is maintained about 25 to 30 centimeters (10-12 inches) below the soil surface during the growing season (Hortus, 1976; Dana, 1990).

New-bog construction is expensive and time consuming. Stump removal, land clearing and leveling, perimeter ditching, and topdressing with several inches of sand (which must be repeated every four to five years) precede scattering and discing-in the cranberry plant cuttings. Hardwood cuttings taken in the early spring and struck into the bed will root rapidly (Cross et al., 1969; Hall, 1969). Rooted cuttings may also be planted on 30 centimeter (12 inch) centers. Seed propagation is possible following three-month stratification (Eck, 1990). The first harvest occurs several years after planting and plantings may persist for many years. The author has visited one bog in Rhode Island that has been harvested more or less continuously since the mid-1700s.

A typical fertilizer regime should aim for soil levels of 65 kilograms/hectare phosphorus and 110 kilograms/hectare potassium with nitrogen applications adjusted to 20 to 30 kilograms/hectare. Base precise application rates on the health and vigor of vegetative growth (Dana, 1990).

Fruit matures 100 to 130 days after bloom. Mature fruit is crisp with the red pigment in the outer tissues only. However, in storage at temperatures above 10°C (50°F) the pigment may diffuse throughout the flesh. Harvesting is done by specialized equipment using dry raking or, more usually, raking or beating the fruit off the plants over a shallowly flooded bog. An average yield is about 100 barrels per acre (1 barrel = 1 hundredweight = 45.359 kilograms). Most of the crop is processed into juice, with about 20 percent going to the fresh market. Sound fruit will keep for two to four months at 3 to 5°C (38-41°F) and 90 to 95 percent relative humidity, though with poor storage ventilation smothering injury may result from low oxygen and high carbon dioxide levels (Lutz and Hardenburg, 1968). The cranberry is processed into a canned sauce, juice, or jelly or may be used fresh or frozen.

cranberry, Alpine *(Vaccinium vitis-idea): See* **lingonberry.**

cranberry, dry ground *(Vaccinium vitis-idea): See* **lingonberry.**

cranberry, highbush: Caprifoliaceae. COWBERRY. About 225 species make up the genus *Viburnum,* the fruit of many of which are edible. Most species are shrubs or small trees native to America, Europe, and Asia. Leaves opposite, simple, deciduous; flowers small, white, in clusters; fruit a persistent drupe. Most species are hardy to zone 4.

Viburnum alnifolium Marsh. AMERICAN WAYFARING TREE, DEVIL'S-SHOESTRINGS, DOGBERRY, DOG-HOBBLE, HOBBLEBERRY, HOBBLEBUSH, MOOSEBERRY, MOOSE BUSH, MOOSEWOOD-HOPPLE, TANGLEFOOT, TANGLE-LEGS, TRIP-TOE, WHITE MOUNTAIN, DOGWOOD, WITCH-HOBBLE, WITCH. The red fruit becomes purplish-black when ripe.

Viburnum lentago L. NANNYBERRY, NANNYPLUM, SHEEPBERRY. This species is a large shrub, up to 8 meters (25 feet) in height, hardy to zone 2 and was introduced in 1761. The elliptical fruit, about 12 millimeters long (0.5 inch) are blue-black and have a heavy, waxy bloom.

Viburnum opulus L. CRANBERRY BUSH, CRAMPBARK, EUROPEAN CRANBERRY BUSH, GUELDERBERRY, GUELDER ROSE, HIGHBUSH CRANBERRY, SNOWBALL TREE, WHITTEN TREE. This species was introduced into Colonial America from Europe and is hardy to zone 3. The barely edible red fruit is extremely tart and about the size of an American cranberry. Maine lumberjacks stewed the fruit with molasses and the berries were eaten with honey in Norway and Sweden. The cultivar 'Xanthocarpum' produces yellow fruit. The bark of this plant has been used medicinally. This species is one of the easiest Viburnums to grow (Hortus, 1976; Hedrick, 1919; Wyman, 1986).

Viburnum prunifolium L. BLACK HAW, STAGBERRY. This large shrub, up to about 4 meters (12 feet) in height, is hardy to zone 3 and was introduced in 1727. The fruit is nearly round, about 12 millimeters in diameter (0.5 inch), blue-black, and covered with a heavy, waxy bloom.

Viburnum trilobum Marsh. CRAMPBARK, CRANBERRY BUSH, CRANBERRY TREE, GROUSEBERRY, HIGHBUSH CRANBERRY, PIMBINA, SQUAWBUSH, SUMMERBERRY, SWEETBERRRY, TREE CRANBERRY. The species is hardy to zone 3 and was introduced into cultivation in 1812. The fruit is similar to that of *V. opulus.*

This species is the one preferred for its fruit, which is used in jellies, pies, and as a general substitute for American cranberries. Unlike fruit of the true cranberry (*Vaccinium* spp.), the fruit of this species contains a single, large seed. It is grown on limited acreage in New Hampshire, Massachusetts, and other northern states of the United States, and also in Canada and northern Europe. Cultivars include 'Andrews', 'Compactum', 'Hahs', 'Manitou', 'Phillips', and 'Wentworth', which are both wind and insect pollinated (Stang, 1990).

Propagation is by stratified seed, hardwood cuttings, or softwood cuttings dipped in 1,000 ppm IBA. The plants are also propagated by layering and grafting (Dirr, 1998). Softwood cuttings rooted in sand or perlite are most successful.

The highbush cranberry grows well on a wide range of soils, preferring those that are a bit on the cool, wet side.

Transplant in the early spring to a site with full sun or light shade, setting the plants about 2 meters (6 feet) apart in rows 5 meters (16 feet) apart. They are not drought tolerant. Apply sufficient irrigation and fertilizer to promote vigorous growth (Patterson, 1957). Maintenance prune to maintain about five canes per bush. Fruit production begins usually in the third season after planting, with full production in the fifth. Fruit is ripe from 90 to 110 days after bloom.

cranberry, hog *(Arctostaphylos uva-ursi): See* **bearberry.**

cranberry, lowbush *(Vaccinium vitis-idea): See* **lingonberry.**

cranberry, moss *(Vaccinium vitis-idea): See* **lingonberry.**

cranberry, mountain *(Vaccinium vitis-idea* L. var. *minus): See* **lingonberry.**

cranberry, rock *(Vaccinium vitis-idea* L. var. *minus): See* **lingonberry.**

creashak *(Arctostaphylos uva-ursi): See* **bearberry.**

creeping barberry: *See* **grape, Oregon.**

creeping mahonia: *See* **grape, Oregon.**

crimsonberry *(Rubus arcticus): See* **blackberry.**

crowberry: Empetraceae. MONOX. The genus *Empetrum* comprises three or four procumbent evergreen shrublets native to northern Eurasia, northern North America, and southern South America. Leaves narrow and whorled; flowers axillary, purplish; fruit a berrylike drupe.
   Empetrum nigrum L. BLACK CROWBERRY, CRAKEBERRY, CURLEW BERRY, MONOX. The plant reaches a height of about 25 centimeters (10 inches) and is hardy to zone 2. Fruit is black and about 0.5 centimeter (0.25 inch) in diameter.

crucifixion berry *(Shepherdia argentea): See* **buffaloberry.**

cup, sweet *(Passiflora maliformis): See* **passionfruit.**

curlewberry *(Empetrum nigrum): See* **crowberry.**

currant: *See Ribes.*

currant, Nebraska *(Shepherdia argentea): See* **buffaloberry.**

currant, zante *(Vitis vinifera): See* **grape.**

curuba *(Passiflora mollissima): See* **passionfruit.**

dangleberry *(Gaylussacia frondosa): See* **huckleberry.**

date, Chinese *(Ziziphus jujuba): See* **jujube.**

date, trebizond: *See* **olive, Russian.**

deerberry: *See* **blueberry.**

desert thorn: *See* **wolfberry.**

devil's-shoestrings *(Viburnum alnifolium):See* **cranberry, highbush.**

dewberry *(Rubus macropetalus): See* **blackberry.**

dogberry: *See* **cranberry, highbush** or **ash, mountain.**

dog-hobble *(Viburnum alnifolium): See* **cranberry, highbush.**

dogwood, white mountain *(Viburnum alnifolium): See* **cranberry, highbush.**

Dovyalis abyssinica (A. Rich.) **Warb:** Flacourtiaceae. This is a bushy shrub up to 3 meters (10 feet) tall, native to Ethiopia and little known outside that country. It is closely related to the **kei-apple** and the **kitembilla** and is excellent for eating out of hand. Leaves are ovate, glabrous, shiny light green, 2.5 to 7.5 centimeters (1-3 inches) long, alternate and lightly armed, or unarmed; monoecious; flowers small, greenish white, inconspicuous, opening mainly September to January. The fruit are oblate, globose, about 2.5 centimeters (1 inch) in diameter and have the color, aroma, and flavor of apricots. One staminate plant per several pistillate plants is a suitable combination for adequate pollination. Plants are subject to nematode infestations and are propagated by seed or grafting (Mowry et al., 1958).

dunks *(Ziziphus mauritiana): See* **jujube.**

eardrops, ladies': *See* **fuchsia berry.**

elaeagnus, autumn *(Elaegnus umbellata): See* **olive, autumn.**

elaeagnus, cherry: *See* **goumi.**

elder: *See* **elderberry.**

elderberry: Caprifoliaceae. About 20 species of shrubs and small trees make up the genus *Sambucus*. All are widely distributed in temperate and subtropic regions. About eight to ten species are native to the United States, though only three to four species are normally used for their edible fruit. Leaves opposite, odd-pinnate; flowers small, white, in cymose clusters; fruit a small (<6 millimeters, <0.25 inch) berrylike drupe borne in a large, flat cluster up to 15 to 23 centimeters (6-9 inches) in diameter.

There is little agreement on species delimitation within this genus, and while the fruit of some species is edible, that of others is poisonous (Hortus, 1976). The genus name is derived from the Latin word *sambuca,* referring to a musical instrument made by boring out the pith of the stem to form a hollow, flutelike instrument. North American Indians and early colonists used the hollow stems as spiles in tapping maple trees, and the tannin in the bark and roots was used to tan leather (Stang, 1990; Ourecky, 1977). An extract also was made from elder flower water and used as a cosmetic. Oil extracted from the seeds of *S. nigra* was used to flavor port wine in Portugal and the flowers flavored German wine. The fruit of edible species is too soft to ship long distances and therefore is used locally in pies, preserves, and wine. The hard, white, fine-grained wood of *S. nigra* takes a high polish and develops a yellow tinge when finished, making it valuable for cabinetwork. The plants were first cultivated in the United States in 1761 (Ourecky, 1977; Ritter and McKee, 1957).

Sambucus caerulea Raf. BLUE ELDER, BLUE ELDERBERRY, BLUEBERRY ELDER, WESTERN ELDERBERRY. This species is found in the wild in the western United States along the Pacific Coast and east to New Mexico and Montana, where it has been cultivated since 1850. The shoots can grow up to 5 meters (16 feet) per year. Bloom occurs from late April to July and the fruit ripens from August to October. The fruit is larger than those of *S. canadensis,* sweet, juicy, blueblack, borne in clusters sometimes weighing several pounds, and used cooked in a manner similar to those of *S. canadensis* (Rehder, 1947). Although the fruit is actually blue-skinned, it appears almost white due to the heavy waxy bloom.

Sambucus callicarpa Greene. PACIFIC COAST RED ELDER. This plant is native from coastal California to Washington and does equally well on well drained soils in partial shade or full sun. The berries ripen from June to September and, in mild autumns, up to December. While the showy scarlet fruit is edible its main use is for browse for sheep and cattle.

Sambucus canadensis L. AMERICAN ELDER, AMERICAN ELDERBERRY, COMMON ELDER, ELDERBERRY, MEXICAN ELDERBERRY, SWEET ELDER, SWEET ELDERBERRY. This shrub grows up to 3 meters (10 feet) in height and is hardy to zone 4. The species is native from Nova Scotia to Minnesota, south to Florida and Texas, and is grown for its purple fruit, which primarily is used for wines, sauces, jellies, and pies. The unopened flower buds, when pickled, are a good substitute for capers (Hedrick, 1919). The leaves, flowers, and fruit have been used for dyes, and the oil from the seeds in medicine and as a flavorant. The plant was introduced into cultivation in the United States in 1761.

Selection and breeding efforts for this genus have been confined mostly to this species. About 508 metric tons (500 tons) of the fruit are produced annually in the United States with limited production in Canada and Eurasia. This species may produce fruit from seed in the first year but more commonly begins to bear in the second or third year. The most productive wood is two years old. Elderberry species in general are considered partially self-fruitful but plant at least two cultivars to be safe. Since the species usually blooms late, blossom damage due to spring frost is uncommon. Some improved cultivars include 'Adams 1' and 'Adams 2' (both self-unfruitful), 'Johns', 'Kent', 'Nova', and 'York'.

The elderberry roots readily from hardwood cuttings about 25 to 46 centimeters (10-18 inches) long (three nodes) made from one-year-old wood in early spring and struck into a rich garden soil so that only the top node is exposed. Rooting occurs quickly and the new plants can be set into their permanent location in autumn.

Elderberries are widely adapted to the United States and suffer little winter damage. They tolerate a wide range of soils from rather moist to rather dry, with the best growth on moist, rich soils with a pH of 5 to 7. Plant in spring in full sun with good air circulation, spacing plants about 2 meters (6 feet) apart in the row and 4 meters (13 feet) between rows. Begin fertilizer applications in the second year with about 28 grams (1 ounce) of actual nitrogen per plant per year of plant age up to a maximum of 225 grams (7.2 ounces) of nitrogen per plant.

The American elder fruits mostly on two-year-old canes so the plants need annual early spring pruning to remove damaged wood and wood older than about three years. Leave about eight vigorous, erect canes per plant and be sure they all arise within a 0.5 meter (2 foot) diameter crown to prevent the bush from "creeping" out of bounds (Ritter and McKee, 1964). Remove all suckers between the rows. Mature plants (after about four years in the field) will produce up to 7 kilograms (18.8 pounds) of fruit. The fruit ripens from 60 to 90 days after bloom and is harvested by picking the entire fruit cluster and stripping the berries off just before processing (Stang, 1990). A healthy, mature plant can yield up to 20 pounds of fruit (Ourecky, 1977).

Undamaged fruit may be held in storage at 1°C (34°F) and 90 to 95 percent relative humidity for 1 to 2 weeks (Lutz and Hardenburg, 1968).

Sambucus ebulis L. DANEWORT, DWARF ELDER, WALLWORT. This is an herbaceous perennial reaching heights of up to 1.3 meters (4 feet). It is native to Eastern Europe but has escaped from cultivation in the United States and now is growing wild in many western states. The fruit is small and black but produces a blue dye.

Sambucus melanocarpa A. Gray. BLACKHEAD ELDER. Plants of this species grow to 4 meters (13 feet) in height. It is native to the west coast of the United States from British Columbia south to California and grows in full sun or partial shade on dry rocky sites or moist sites. Although the fruit is edible, it is primarily used as forage for wildlife. The plant is hardy to zone 6.

Sambucus nigra L. EUROPEAN ELDERBERRY. This plant is hardy to zone 6 and native to Europe, North Africa, and western Asia. It has escaped from cultivation in the United States. The plant grows to 8 meters (26 feet) and produces heavily fragrant flowers and small, black, edible fruit. The wood is used in cabinetmaking.

Sambucus pubens Michx. AMERICAN RED ELDERBERRY, RED-BERRIED ELDER, STINKING ELDER. This moderately large shrub grows to 5 meters (16 feet) and produces pinkish-white flowers and blackish-red fruit. It grows throughout the United States and is tolerant to either shade or full sun, well drained or moist soils. The fruit is sometimes considered poisonous to humans, but evidence for this varies and there are numerous reports of its use in cooking. It is a favorite food of wildlife.

Sambucus racemosa L. AMERICAN RED ELDER, EUROPEAN EL-DER, EUROPEAN RED ELDER, RED ELDER, RED-BERRIED ELDER, SCARLET ELDER, STINKING ELDER. This rather tall bush, growing up to 4 meters (13 feet), is native to Europe and western Asia and has escaped from cultivation in the United States, into which it was introduced in 1596. Its fruit is scarlet as compared to the blackish-red fruit of *S. pubens*. Some sources list *S. pubens* as a subspecies of *S. racemosa*. The new shoots of this species are distinctly reddish and usually have four ridges around the stem, while those of *S. pubens* are glaucous (Ritter and McKee, 1964).

encarnado: *See* **lingonberry.**

Eugenia: Myrtaceae. More than 1,000 species make up the genus *Eugenia*. All are evergreen trees or shrubs native mostly to the American tropics. Leaves opposite, simple; flowers mostly in clusters; fruit a one or two seeded berry and sometimes edible.

Eugenia aggregata (Vell.) Kiaersk. CHERRY-OF-THE-RIO-GRANDE. The cherry-flavored fruit is an oblong berry about 2.5 centimeters (1 inch) in length turning from orange red to deep purple red.

Eugenia luschnathiana Klotzsch ex O. Berg. PITOMBA. A shrub grown in Brazil for its soft, succulent, aromatic fruit.

Eugenia pitanga (O. Berg) Kiaersk. PITANGA. A low shrub grown in Brazil and Argentina for its fruit, which makes a good jelly.

Eugenia uniflora L. BARBADOS CHERRY, BRAZIL CHERRY, CAY-ENNE CHERRY, PITANGA, SURINAM CHERRY. This plant is hardy to zone 10b and forms a large shrub or small tree widely cultivated in the tropics as an ornamental, and for its 2.5 centimeter (1 inch) diameter ridged, globose fruit, which is used in jellies, jams, relish, pickles, and sherbets (Hortus, 1976; Wyman, 1986). Each fruit contains one or two seeds and ripens 30 to 50 days after bloom. The flavor is tart, but somewhat less tart than that of a tart cherry *(Prunus cerasus).* Production areas include Hawaii, Florida, the West Indies, and South America. In the tropics, two crops per year are produced, sometimes commercially. In the continental United States, one crop per year is produced on a noncommercial basis. The plant is propagated by seed or by cuttings.

Feijoa: Myrtaceae. This family contains two species of evergreen shrubs native to South America. Leaves opposite; flowers single, axillary; fruit a lobed, oblong berry 3 to 7 centimeters (1.2-2.75 inches) long with persistent calyx.

Feijoa sellowiana O. Berg. PINEAPPLE GUAVA. This is the only species grown for its fruit, which is green tinged with red and about 8 centimeters (3 inches) in length. It is hardy to zone 9 but tolerates temperatures as low as –9.5°C (15°F). Although also tolerant to drought and saline soils, the plants produce best where annual rainfall is 1.5 meters (5 feet). Commercial production in the United States occurs mainly in California, where it was introduced around 1900, with smaller production areas in Florida and Hawaii. There are probably less than 1,000 acres grown domestically. The fruit is also grown in Brazil, Paraguay, Uruguay, northern Argentina, and New Zealand.

The culture of this fruit is similar to that of the true guava (Hortus, 1976; Wyman, 1986). Cultivars include 'Andre' and 'Coolidge', both self-fruitful, and 'Triumph', 'Choiciana', and 'Superba', all of which require cross-pollination to set good crops. Bees are the chief means of pollination and fruit set approximates 60 to 90 percent under normal conditions. The plants are propagated from cuttings, layering, or grafting onto their own stock by whip, tongue, or veneer grafts.

Plants do best on rich, well drained loam soils at spacings of 5 to 6 meters (16-20 feet) between rows and 3 to 4 meters (10-13 feet) between plants within rows. They tolerate partial shade and develop shallow, fibrous root systems. All plants are trained to a single stem with 4 to 5 branches and are fertilized as citrus. The first crop is produced two to three years after planting.

Fruit is harvested 150 to 180 days after bloom when it turns light green and fall to the ground. The edible seeds and the pulp make fine jelly, marmalade and other preserves, with a taste resembling guava (Mowry et al., 1958). About 4 to 10 metric tons (4-11 tons) per acre are produced in California, though worldwide the yield per plant ranges from 100 fruits (India) to 2,000 fruits (Riviera) (Morton,

1987). Fruit will keep for approximately one month when stored below 10°C (50°F).

fig, Barbary *(Opuntia vulgaris): See* **pear, prickly.**

fig, barberry *(Opuntia vulgaris): See* **pear, prickly**.

fig, bastard *(Opuntia phaeacantha): See* **pear, prickly.**

fig, Hottentot: Aizoaceae. About 29 species of subshrubs make up the genus *Carpobrotus*. All are native to Africa, Australia, Tasmania, and North and South America and many have become naturalized in Europe. Two species carry the common name HOTTENTOT FIG. Leaves opposite, three-angled; flowers solitary; fruit indehiscent and pulpy. The plants make an excellent sand cover along the coast (Wyman, 1986).

Carpobrotus acinaciformis (L.) L. Bolus. This trailing species reaches heights of about 2 meters (6.5 feet) and has large purple flowers about 15 centimeters (6 inches) in diameter that open at noon.

Carpobrotus edulis (L.) L. Bolus. Plants of this trailing species reach about 1 meter (3 feet) in height. Their bright yellow flowers also open at noon. This species escapes from cultivation especially easily and can become problematic, as in southern California (Hortus, 1976). The fruit flesh makes a good preserve and the leaves of the plant can be pickled as a substitute for cucumber pickles. The plant is hardy to zone 10, grows well in rich soil, and is easily propagated by cuttings.

fig, Indian *(Opuntia ficus-indica): See* **pear, prickly.**

filbert: Betulaceae. COBNUT. The genus *Corylus* contains about 18 species of deciduous, monoecious shrubs or small trees native to northern temperate regions. Leaves alternate; flowers unisexual, male flowers in catkins; fruit a nut. While the fruit of all species is edible, two species, *C. avellana* and *C. maxima,* are the most important cultivated filberts. The names "filbert" and "hazelnut" are used interchangeably (Woodroof, 1979).

Corylus americana Marshall. AMERICAN FILBERT, AMERICAN HA-
ZELNUT. This plant is native to the eastern United States and Canada,
where the nuts are harvested locally. The species has been used in
breeding programs to confer cold hardiness and disease resistance to
hybrid progeny (Hummer, 1999).

Corylus avellana L. AVELLANA, EUROPEAN FILBERT, EUROPEAN
HAZELNUT. Plants of this species native to Europe reach heights of 8
meters (26 feet) and produce large nuts commercially in Europe. The
nut size and quality exceeds those in other *Corylus* species. The
cultivar Grandis is sometimes called the COBNUT.

Corylus cornuta var. *californica* (A. DC.) W.M. Sharp. CALIFOR-
NIA HAZELNUT, BEAKED FILBERT, BEAKED HAZELNUT, WESTERN
HAZELNUT. This species is native to the western United States and
produces relatively low-quality nuts harvested locally.

Corylus heterophylla. SIBERIAN HAZELNUT. This species is native
to China, Korea, and Japan and bears good quality nuts harvested locally.

Corylus maxima Mill. GIANT FILBERT. Plants of this species reach
up to 10 meters (33 feet) in height and produce large nuts used commercially. The nuts are nearly completely enclosed in involucres but
their apex is partially exposed. The shell is hard and woody and the
kernel lies free. Many commercial cultivars used today are hybrids of
C. avellana × *C. maxima*. Commercial production is limited to four
regions of the world, all in close proximity to large bodies of water.
Turkey led the world in 2004, producing 425,000 metric tons
(468,482 tons), followed by Italy with 135,000 metric tons (148,812
tons), and the United States, with 48,500 metric tons (53,462 tons)
(http://www.fas.usda.gov). The Willamette Valley of Oregon produced 99 percent of the U.S. crop in 1992 (Hummer, 1999), with the
remaining acreage in Washington (Markle et al., 1998).

In general, the filbert catkins, even when fully dormant, are killed
by temperatures below about −10°C (15°F). The European species is
not suitable for planting in the eastern United States because of its
lack of hardiness and because of its susceptibility to *Cryptosporella
anomala* (C.H. Peck) Cacc., a fungus blight. Therefore, *C. avellana*
has been hybridized with two native species, *C. cornuta* and *C.
Americana,* to produce cultivars suitable for the region. 'Bixby', 'Bu-
chanan' (or 'Buchannon'), and 'Potomac' are three such cultivars. A

bacterial blight, *Xanthomonas corylina* (P.W. Mill. et al.) M.P. Starr & Burkholder, can be a problem on young trees in Oregon.

Historical use of the nuts is ancient. The Chinese used them more than 5,000 years ago (Peker, 1962). Ancient Greeks and Romans used the nuts for both medicine and food. In the first century AD, Dioscorides reported the nuts cured the common cold as well as baldness. In Roman and Celtic folklore the hazelnut was thought to have mystical powers to protect against lightning strikes and to divine the location of treasures and water (Hummer, 1995). The nutmeats are quite high in protein and the shells have been used in making plywood and linoleum. The very dense smoke produced when the shells are burned was once used in production of poisonous gases and gas masks (Woodroof, 1979).

The pistillate flowers are borne in lateral and terminal buds on one-year-old wood. The staminate flowers are borne on short stalks laterally on one-year-old wood or terminally on short spurs. Bloom occurs from midwinter to early spring. This species is mostly dichogamous, self-unfruitful, and wind-pollinated. This self-sterility makes interplanting necessary for good cross-pollination, with the proportion of one pollinizing tree for each eight trees of the main cultivar appropriate.

The filbert industry is mostly based upon the single cultivar 'Barcelona', with another important cultivar being 'Du Chilly'. Since both are self-sterile, a third cultivar, 'Daviana', is used as a pollinizer. Plants are propagated by mound and tip layering (Hartmann et al., 2002). Plant filberts in deep, rich, well drained loam soils with a pH of about 6, on sites with good air and water drainage. Set the plants about 7 meters (22 feet) on center and mound soil or sawdust over their crowns to reduce suckering (Sander, 1963). Although the plants are naturally bushes, allowing the suckers to form freely will make management difficult. Instead, remove the suckers as they form and train to a small tree shape. Head newly planted trees at about 0.6 meter (2 feet) and protect the trunks from sunscald by wrapping or painting. Good weed control is very important. The plants begin to bear about four to six years after planting since most fruit is produced on new wood. It is important to keep the tree producing a generous supply of new wood through proper pruning and fertilizing. Shoot growth should average 15 centimeters (6 inches) per year. One method of pruning in Oregon is to remove half the fruiting wood from one fifth of the trees each year (Olsen, undated). Fertilizer rates, of course, depend upon location, age of planting, etc. Oregon recommends a mature tree receive 0.5 to 0.75 kilogram (1.5-2 pounds) of actual nitrogen per year, broadcast beneath the tree's dripline.

Filberts are harvested after they have matured and fallen, or about four months after leaves appear. They are cleaned mechanically or by hand. A good yield is around 750 kilograms (1,653 pounds) per acre. The nuts must be gathered within a week or so after they drop so as to reduce shell discoloration. Dry the nuts within a week or so by placing them in an oven at about 32 to 38°C (90-100°F) until their moisture content is reduced to 8 to 10 percent. Nuts at that moisture will snap if bitten when cold.

foxberry *(Vaccinium vitis-idea): See* **lingonberry.**

fraises des bois *(Fragaria vesca): See* **strawberry.**

framboise *(Rubus idaeus): See* **raspberry.**

fruitillar *(Fragaria chiloensis): See* **strawberry.**

fruit-salad plant *(Monstera deliciosa): See* **monstera.**

fuchsia berry: Onagraceae. About 100 species of shrubs and trees make up the genus *Fuchsia*, commonly called FUCHSIA, LADY'S EAR-DROPS, or LADIES' EARDROPS. They are native from Mexico to southern South America and also in New Zealand and Tahiti. Leaves alternate, opposite or whorled; flowers showy red to purple and white; fruit a berry (Hortus, 1976).

Fuchsia corymbiflora Ruiz & Pav. This shrub grows up to 5 meters (16 feet) in height and is native to Ecuador and Peru, producing edible fruit harvested from the wild.

Fuchsia denticulata Ruiz & Pav. This bush up to 3 meters (10 feet) in height grows wild in Peru and Bolivia, producing edible, though acid, fruit (Hedrick, 1919).

Both of these species also produce edible flowers and the plants could probably be propagated by leafy cuttings under mist (Wallis, 1976).

G

gayuba *(Arctostaphyllos uva-ursi):* See **bearberry.**

goldenberry *(Physalis peruviana):* See **cherry, ground.**

gooseberry: See *Ribes.*

gooseberry, cape *(Physalis peruviana):* See **cherry, ground.**

gooseberry, Chinese *(Actinidia chinensis):* See **kiwifruit.**

gooseberry, dwarf cape *(Physalis pruinosa):* See **cherry, ground.**

gophla *(Holboella latifolia):* See *Holboella.*

goumi/gumi: Elaeagnaceae. CHERRY ELAEAGNUS. This plant, *Elaeagnus multiflora* Thunb., is a close relative of buffaloberry and forms a deciduous bush up to 3 meters (10 feet) in height. It is native to China and Japan, from whence it was introduced into the United States by Commodore Perry in 1862. The plant has silvery leaves and bears pleasant-tasting, slightly acid scarlet fruit about 1 centimeter (0.4 inch) long in midsummer. The plant is hardy to zone 4 (Hortus, 1976; Hedrick, 1919; Wyman, 1986). This species may be propagated by stratifying seeds for four months at 4°C (40°F). Both hardwood and softwood cuttings of related species root well (Hartmann et al., 2002).

governor's plum: Flacourtiaceae. RAMONTCHI, RUKAM. *Flacourtia indica* Merr. is a shrub native to Madagascar and southern Asia adapted to areas of the United States that experience little frost. General height is about 3.5 meters (12 feet) and the bush is usually lightly armed. Leaves 0.6 to 1 meter (2-3 feet) long, ovate to oblong obovate, with dentateserrate edges, glabrous, deep green above and slight paler below, glossy and slightly leathery; fruit subglobose berry about 2.5 centimeters (1 inch) in diameter, maturing in summer; fruit skin almost black, flesh juicy and ranges from sweet to acid. The plants are dioecious. This species is propagated by seed, cuttings of mature wood, or by grafting. The culture of this entire genus is easy and plants grow well on both sandy and limey soils. Prune to keep plants in bounds (Mowry et al., 1958).

granadilla *(Passiflora edulis): See* **passionfruit.**

granadilla, giant *(Passiflora quadrangularis): See* **passionfruit.**

granadilla, sweet *(Passiflora ligularis): See* **passionfruit.**

grape: Vitaceae. This family contains 12 genera distributed throughout the world. Grape belongs to the genus *Vitis,* itself widely distributed throughout the northern hemisphere. All *Vitis* have exfoliating bark, are woody vines, and climb by means of tendrils. Leaves simple; flowers small, unisexual; fruit an ovoid berry with soft pulp and often borne in clusters. While there are a few dozen species within this genus, only very few, along with their hybrids, are grown commercially (Hortus, 1976).

Vitis labrusca L. AMERICAN BUNCH GRAPE, FOX GRAPE, SKUNK GRAPE, SLIPSKIN GRAPE, PARSON. The fruit is borne in clusters. The berries are about 2 centimeters (0.75 inch) in diameter, purple-black, red-brown, red, or green in color, sweet, with a strong musky ("foxy") flavor and aroma as found in commercially prepared grape juices, jams, and jellies. The skin is thin and waxy and the pulp, along with two to four seeds, separates from the skin at maturity. This species is more winter hardy and disease tolerant than the others in commercial production but lacks the delicate flavor of *V. vinifera.* It is hardy to zone 5. This species and its many hybrids and cultivars are the mainstay of the U.S. grape industry east of the Mississippi River, with the largest plantings located on the shores of the Great Lakes. The species does best where summer humidity is moderate and is not well suited to arid regions. Some cultivars are grown in every state in the continental United States.

Vitis rotundifolia Michx. BULLACE, BULLACE GRAPE, MUSCADINE GRAPE, SCUPPERNONG, SOUTHERN FOX GRAPE. The fruit of this grape is borne singly or in small clusters with the dull-purple, thick-skinned berries about 2.5 centimeters (1 inch) in diameter. The skin separates from the pulp. This was the first American species to be cultivated and the fruit has a strong, musky-flavored pulp. Cultivars can be divided into wine types, table types, and juice/jam types. The vines are vigorous and resistant to many diseases but are intolerant of temperatures below –18°C (0°F). In fact, they thrive only in the warm, humid areas of the southeastern United States, being hardy only to zone 6 (Winkler, 1974).

Vitis vinifera L. CHAMPAGNE GRAPE, CORINTHIAN GRAPE, CURRANT GRAPE, EUROPEAN GRAPE, OLD WORLD GRAPE, WINE GRAPE, ZANTE CURRANT. Fruit of this species is cultivated in many forms throughout the world and in the United States, mostly in California and Arizona. The fruit is borne in clusters with berries up to 2.5 centimeters (1 inch) in diameter. Fruit skin is light green to black or red, smooth and waxy, and remains attached to the pulp at maturity. This grape is the foundation of the wine industry and its flavor and aroma are characterized as "vinous." The species also largely contributes to the table grape and raisin industries. Some pea-sized berries are sold as "champagne grapes" when fresh and Zante currants when dried. Most cultivars lack winter hardiness and are highly susceptible to diseases. They are hardy to zone 5b and require long, warm-to-hot, dry summers and cool winters (Winkler, 1974).

In order to increase cold hardiness and disease tolerance, French researchers hybridized this species extensively with some American species, such as *V. labrusca*. This group of hybrids is termed "French hybrids" (Hortus, 1976).

In 2004, Italy led the world in grape production with 8,691,970 metric tons (9,581,256 tons), followed by France with 7,542,036 metric tons (8,313,671 tons) and Spain with 7,147,000 metric tons (7,878,218 tons) (http://www.faostat.fao.org). The United States produced 5,418,160 metric tons (5,972,499 tons). California led U.S. production with 5,360,000 tons (4,862,510 metric tons), followed by Washington with 267,000 tons (242,218 metric tons) and New York with 145,000 tons (131,541 metric tons) (http://www.jan.mannlib.cornell.edu). Most grapes produced in California were for wine production (2,700,000 tons or 2,449,398 metric tons) followed by those for raisins (1,930,000 tons or 1,750,866 metric tons) and for table use (743,000 tons or 674,038 metric tons).

Grapes are thought to have originated around the Caspian Sea and their virtues were extolled in sacred writings. The vinifera grapes have the oldest history of use of any, with Egyptian records on winemaking dating back perhaps 6,000 years. Winemaking in ancient Greece and Rome was an important part of the economy. The grape was first introduced into England either by the Romans or by the

Phoenicians. However, the climate in England is not suitable for grape culture and the vines were grown primarily under glass. The vinifera grape was introduced into the New World by Columbus. While the early American colonists harvested wild grapes, they were persistent in trying to cultivate vinifera types. This failed in most of the United States except in California, where the Mission Fathers moving northward from Mexico established the San Diego mission in 1769. It was there and in other missions that they planted the grape stock they had brought with them (Winkler et al., 1974; Snyder, 1937; Hedrick, 1919).

There are many species of grapes grown around the world, but all can be divided into five main classes according to their commercial use (Winkler et al., 1974; Jacob, 1959). Table types are used for food and decorative purposes and must have an attractive appearance, good eating qualities, good shipping and storage qualities, and resistance to handling injury. Usually this includes large berries with firm pulp and tough skin. There is a strong preference in the United States for seedless grapes. Most standard table cultivars are *V. vinifera*. Wine grapes can be cultivars of any species, but most of the finer wines are made from *V. vinifera* or from French hybrids. Dry (table) wines are made from grapes with high acidity and moderate sugar while dessert wines are made from fruit of low acidity and high sugar. Raisin grapes include any dried grape that makes a finished product (raisins) wherein the individual dried fruit is soft and will not stick together when stored. Good flavored, seedless cultivars are preferred, commonly 'Thompson Seedless' and 'Muscat of Alexandria'. Juice grapes should be such that the processing does not destroy the natural flavor of the grape. Much of the flavor of the vinifera cultivars is destroyed during processing but the labrusca cultivars manage to maintain their "foxiness" and so are preferred, with 'Concord' the standard juice grape in the United States. Canning grapes are seedless, with 'Thompson Seedless' the most commonly used cultivar.

The number of cultivars within each species is legion (there are at least 5,000 named cultivars of *V. vinifera* alone) and their selection is often predicated upon their tolerance to cold. Only the most cold tolerant cultivars are grown where winter lows reach –29°C (–20°F). These are cultivars such as 'Beta', 'Blue Jay', 'Red Amber', 'Swenson Red', and 'Valiant', that acquired their cold tolerance from native

North American species such as *V. riparia,* the riverbank grape. In the more temperate regions of northeastern and central North America, 'Concord', 'Catawba', 'Delaware' (one of the highest quality American-type grapes), and 'Niagara' (the leading American-type white grape) are reliable producers, along with 'Steuban', 'Ontario', 'Worden', and 'Fredonia'. American bunch grapes for milder regions include 'Champanel', 'Lenoir', 'Portland', and 'Fredonia'. 'Stover', 'Blue Lake', and 'Lake Emerald' are reportedly adapted to Florida conditions. Standard muscadine cultivars are 'Hunt', 'Thomas', and 'Scuppernong', although some new cultivars, such as 'Fry', 'Higgans', and 'Magnolia' (all bronze-skinned) and 'Albermarle', 'Cowart', and 'Magoon' (black-skinned cultivars) show some promise.

Cultivars of *V. vinifera* grown commercially on the west coast of the United States for table grapes include 'Cardinal', 'Thompson Seedless', and 'Flame Tokay'. Important wine cultivars include 'Pinot Noir', 'Zinfandel', and 'Cabernet Sauvignon' (reds) and 'Chardonnay', 'Muscat of Alexandria', and 'White Riesling' (whites) (Ahmedulla, 1996). French hybrid cultivars include the white cultivars 'Aurore' and 'Seyval' and the red cultivars 'Baco Noir' and 'Kuhlmann 188-2' (also known as 'Marechal Foch'). Many French hybrid cultivars carry the name of the breeder and the seedling or selection number. The flavor of French hybrid fruit is usually more neutral than that of any of its parents (Einset, 1973).

There is a difference of opinion as to whether the grape is self-pollinated or pollinated by wind or insects. No doubt all means are effective, but there is growing consensus that the grape is primarily self-pollinated (Winkler et al., 1974).

Cultivars of French hybrids, *V. labrusca,* and *V. vinifera* can be propagated by hardwood cuttings containing three nodes, taken in late winter, and callused in moist sand under cool conditions. The callused cuttings are planted out in the early spring with only their top node exposed above the soil line. The cuttings root rapidly and form usable vines in their second year. The roots of many *V. vinifera* are susceptible to the root louse or grape phylloxera *(Phylloxera vitifoliae)* and also to several species of the root-knot nematode *(Meloidogyne* spp.), and so these vines are usually grafted to a rootstock resistant to these pests. Such rootstocks include species of grapes native to the Mississippi Valley region, such as *V. riparia, V.*

rupestris, V. candicans, V. champini, and *V. rufotomentosa* (Kunde et al., 1968). Field budding (chip budding), bench grafting, and field grafting in spring using whip, cleft, notch, and bark grafts are acceptable means of forming vines on various rootstocks (Weaver, 1976). Cultivars of *V. rotundifolia* and other species are propagated by layering or by softwood cuttings. Vines propagated from seed will not be true to type.

Grapes are deep-rooted and do well on a deep (at least 1.3 meters, 4 feet), friable, well drained, sandy-loam soil with slightly acid to neutral pH (Snyder, 1937). The deeper and more fertile soils usually produce the heaviest crops and so are preferred for production of table, raisin, and standard wine grapes (Weaver, 1976). Excessively fertile soils promote excess vegetative growth that is subject to winter kill and elevated pest attack. Vineyards should be located in full sun in areas with an average of 180 and a minimum of 155 frost-free days. The warm temperate zone, between 34° and 49° north and south latitudes, is the most successful grape growing region, especially where the mean temperature of the warmest month is in excess of 19°C (66°F) and of the coldest month in excess of –1°C (30°F) (Winkler et al., 1974; Prescott, 1965). Excessively cool weather often results in low sugar/high acidity fruit. A south-facing slope speeds fruit ripening and hastens harvest. Of greater importance than the number of frost-free days is the number of heat units accumulated during a typical season.

Good air drainage is necessary to allow cold air to flow away from the vines. Since grapes bloom fairly late, spring frosts are not often a problem. However, early fall frosts may abbreviate the ripening season unless sufficient air drainage is provided.

Planting is usually done in early spring using certified, diseasefree one- or two-year-old vines. Vine spacing varies according to species, region, and cultivation practices (Weaver, 1976). In the northeastern United States typical spacings are 8 feet apart within the row spaced about 9 feet between rows. *Vitis rotundifolia* vines are usually spaced about 10 to 18 feet apart within rows spaced about 10 feet apart. The spacing between rows depends largely upon the type of equipment used in vineyard management. Of primary consideration in determining row direction is the contour of the land. In hilly land, cross-slope planting is necessary to minimize soil erosion. Where land is not sloped, orient the rows east/west to provide for good sunlight penetration into the canopy though the lack of air drainage may become problematic. Set vines on their own roots so that only two to three buds remain above the soil line after pruning. Set grafted vines so that the graft union remains about 9 centimeters (3.5 inches) above the soil on level ground (Weaver, 1976). Erect the trellis or stake system soon after planting and cut the vine back to two to three buds at planting. After growth begins, tie the strongest shoot to the stake or trellis for training as the trunk of the future plant.

The precise system of training the vines depends upon which type of support you have erected, and hundreds of types have been developed over the years. The following list provides descriptions of a few of the more common types:

Four-arm kniffen: This is perhaps the most popular training system for home and small commercial production. Trellis construction involves erection of two wires, the first about 1 meter (3 feet) off the ground and the second 0.6 meter (2 feet) above the first. Wires are held by posts spaced about 2.6 meters (8.5 feet) apart. The trunk is trained to the top wire and four short arms, two on each side of the trunk close to the lower and upper wires, are selected. Canes arising from the arms are trained along the wires, one per wire per side of the trunk. Top canes are more productive than bottom canes.

Six-arm kniffen: This is similar to the four-arm kniffen except that it employs three wires and six arms and canes rather than four.

Umbrella kniffen: This system also employs four canes, all of which are draped over the top wire and then allowed to fall to the bottom wire, to which they are tied.

Hudson River umbrella: This appears at first to be similar to an umbrella kniffen but actually employs two trunks (as insurance against winter kill) (Shaulis et al., 1968). Each trunk is trained to the top of a two-wire system similar to the four-arm kniffen and allowed to extend along the upper wire, one in each direction away from the trunk. Three long canes are allowed to drape to the lower wire.

The grape vine has its own specialized vocabulary of terms, most of which are not used in reference to other crops. It would be well to understand the following terms before a discussion of pruning and training (Winkler et al., 1974; Jacob, 1959):

Arm: A branch older than one year.

Fruit cane: The basal part of a mature cane about 12 buds long that produces fruit the following season and is removed in pruning early the following spring.

Half-canes or rods: Canes cut back to five to ten buds.

Spur: Basal portion of the cane, from one to four buds long.

Renewal spur: This produces a cane to be used as a fruit cane the following season.

Replacement spur: Used to replace arms or branches and usually produce fruit.

Suckers: Properly, a vigorous shoot arising from below the soil surface. More commonly, any vigorous shoot arising from either below ground or from the trunk.

Trunk: The main stem of the vine.

In the early spring one year after planting, prune the vine to one or two buds on the strongest cane and remove all other canes and suckers. In the second spring after planting, head back the trunk cane to force strong branching and tie these shoots to the trellis system as required. In the third spring after planting, select two shoots growing in opposite directions from the trunk for each wire of the system and head back the shoots to five to seven buds to prevent early overbearing. Continue to prune your grapes in early spring annually, remembering that the grape plant is the most severely pruned of all the commonly grown small fruit, with over 90 percent of the wood removed each year to maintain productivity. A full discussion of grape pruning is beyond the scope of this book. Consult any of several other works for details on pruning this plant (Winkler et al., 1974; Weaver, 1976).

Adequate irrigation is necessary for good production, with drip irrigation generally preferred to overhead irrigation since it keeps the leaf and shoot surfaces dry and therefore less subject to diseases. The correct amount of supplemental irrigation required depends upon age of the vine, soil type, atmospheric conditions, vineyard location and region, cultivar, stage of maturity of the crop, etc. but should at least equal the moisture depletion due to evapotranspiration.

As in other fruit crops, nitrogen is the nutrient most likely to be deficient in the vineyard. Symptoms of nitrogen deficiency include yellowing of the older leaves, light green foliage, and reduced shoot growth. Yields may be reduced before a deficiency in this nutrient becomes apparent. Deficiencies in other nutrients, particularly iron, potassium, and magnesium, may also occur in some locations under some conditions. An iron deficiency appears as interveinal chlorosis on the newest leaves with symptoms eventually spreading to older leaves. Potassium-deficient vines display marginal necrosis of the leaves, with general chlorosis developing and the leaf eventually taking on a purple cast. A deficiency in magnesium appears as a red or yellow interveinal chlorosis, with the oldest leaves being the first to show the symptoms. Take measures to correct all of these deficiencies only after they have been properly diagnosed (Weaver, 1976).

Because so many species and cultivars of grapes are grown over such a diverse area under varying conditions it is not possible to make a blanket fertilizer recommendation here. Consult your local cooperative extension service for fertilizer recommendations for your area.

Grapes will not ripen and color fully if harvested immature. Since red and black grapes develop their color slowly as they ripen, a fully colored grape is often the ripest. Green or white grapes take on a yellowish tinge when ripe. Other indicators include development of characteristic flavor and aroma, browning of the peduncle (cluster stem) and pedicel (berry stem), browning of the seeds, and thickening of the juice from watery to syrupy (Weaver, 1976). Harvest the fruit by snipping the peduncles from the shoot, being careful not to handle the fruit too much and so remove the delicate bloom. *V. vinifera* grapes can be kept in good condition for up to ten days by holding them at around 1°C (34°F) and 90 to 95 percent relative humidity. *V. labrusca* and muscadine grapes can be held at the same temperature but at a slightly lower relative humidity of 85 percent. *V. labrusca* grapes keep for up to a month, while the muscadines remain in good condition for only three weeks or so. Berries stored too long lose their brightness and become flaccid (Lutz and Hardenburg, 1968).

grape, bear's *(Arctostaphylos uva-ursi): See* **bearberry.**

grape, mountain *(Mahonia aquifolium): See* **grape, Oregon.**

grape, Oregon: Berberidaceae. CREEPING BARBERRY, CREEPING MAHONIA. This is not a true grape, and while this fruit shares a family with barberry, it has recently been assigned its own genus, *Mahonia*. This genus contains more than 100 species of evergreen, thornless shrubs native to North and Central America and to Asia. Leaves alternate, odd-pinnate; leaflets mostly spiny; flowers yellow, usually in clusters; fruit a dark blue, strongly acid berry, with bloom (Hortus, 1976). Many species bear edible fruit.

   Mahonia aquifolium. BLUE BARBERRY, HOLLY BARBERRY, HOLLY-GRAPE, HOLLY MAHONIA, MOUNTAIN GRAPE. Most commonly it is called OREGON GRAPE, though this name is applied to many species in this genus. The plant entered into cultivation around 1900 (Rehder, 1947). Prune the plants to about 1 meter (3 feet) in height since, if allowed to grow taller they sometimes flop and become unsightly with loss of fruit production (Wyman, 1986). The fruit is sometimes fermented with added sugar, making an acceptable wine. They are also used in confections and sometimes eaten for their antiscorbutic properties (Hedrick, 1919). Some species of *Mahonia* are hardy to zone 5 but must be protected from wind and bright sun, doing best in moist shade. Propagation is by seed, sucker, layers, and hardwood and semihardwood cuttings. Seed of *M. aquifolium* should be separated from the pulp and leached for a week or so under continuous-flow water and stratified for five months (Hartmann et al., 2002). Collect hardwood cuttings in fall or winter, dip in 3000 ppm IBA, and root under mist with 25°C (77°F) bottom heat. Different species respond differently to treatments (Dirr and Heuser, 1987).

grape-pear *(Amelanchier canadensis): See* **serviceberry.**

grouseberry: *See* **blueberry** or **cranberry, highbush.**

guava: Myrtaceae. CAS, SAND PLUM. This family contains about 100 species of trees and shrubs native to the American tropics, several of which are cultivated for their edible fruit. Leaves simple, opposite; flowers white, large; fruit a globose or pear-shaped berry.

   Psidium acutangulum DC. PARA GUAVA. This plant forms a shrub or tree 8 to 12 meters (26-40 feet) in height. Branches quadrangular and winged near the base of leaves. Leaves elliptical, 10 to 14 centimeters (4-5 inches) long, 4 to 6 centimeters (1.5-2.4 inches) wide. White, five-petalled flowers. Fruit round to ellipsoid, 3 to 8 centimeters (1.2-3.25 inches) wide, pale yellow, very acid, pleasant flavor. The plant grows wild and is cultivated in Amazonia and throughout much of South America, and is sometimes grown in southern Florida under the name *P. araca.* The fruit is eaten with honey or made into acid drinks or preserves.

Psidium friedrichsthalaianum (O. Berg) Niedenzu. COSTA RICAN GUAVA. This white-fleshed, sulfur-yellow fruit up to 7 centimeters (2.75 inches) long is smaller than that of the common guava and has an agreeably tart flavor. It is most often used for jellies.

Psidium guajava L. APPLE GUAVA, COMMON GUAVA, LEMON GUAVA, YELLOW GUAVA. This species is cultivated throughout the tropics and subtropics and has naturalized over much of its area, becoming a weed in some places. The fruit varies in color and shape and is too strongly flavored when eaten raw. Hence, it is often canned, spiced, and processed into jams, chutney, and, most commercially important, jellies. Guava juice is used in punch. In the United States this species is usually grown in Hawaii and subtropical Florida, and is hardy to zone 10b.

Psidium guineense Swartz. BRAZILIAN GUAVA, CASTILIAN GUAVA, GUAVA. This evergreen shrub grows to a height of 1 to 3 meters (3-10 feet) and sometimes forms a tree to 7 meters (23 feet). The fruit is small, round or pear-shaped, about 1 to 2.5 centimeters (0.12-1 inch) diameter, and has yellow skin and thick, pale yellow flesh. It is resinous and acid and, like *P. guajava,* is too strongly flavored to be palatable when eaten raw. It is the most widely distributed guava but is presently under limited cultivation in Martinique, Guadeloupe, the Dominican Republic, and southern California. It has become naturalized in northeastern India. There are no named cultivars, but this species when crossed with the common guava produces a dwarf, hardy, highly productive hybrid. The plant does best on good, rich soil and the fruit is baked and used in preserves, making a distinctive jelly. The wood is strong and used for handles, beams, and other construction purposes. The high tannin content of the bark makes it useful for tanning hides. In interior Brazil the bark is used to treat urinary diseases and dysentery, while it is used in Costa Rica to treat varicose veins and ulcers, colds, and bronchitis (Morton, 1987).

Psidium littorale Raddi. YELLOW CATTLEY GUAVA, YELLOW STRAWBERRY GUAVA, WAIAWI. This species is also cultivated over a wide area of the tropics and subtropics, primarily for reforestation, although its yellow fruit is edible. It is hardy to zone 10.

Psidium littorale var. *longipes.* CATTLEY GUAVA, PURPLE GUAVA, PURPLE STRAWBERRY GUAVA. The purple-red sweet, soft fruit is quite edible and make this often the preferred species among the guavas. It is about as hardy as lemon and in the United States is grown in southern California and Florida (Hortus, 1976). While there was extensive commercial production in the United States around 1900 this production has substantially declined over the last century. In 1995, Florida produced about 60 hectares (150 acres) and Hawaii about 300 hectares (750 acres) (Markle et al., 1998). Guam and Puerto Rico are smaller producers. Other areas of world production include India, Egypt, Brazil, Mexico, and South Africa. Guava was carried into the East Indies by the Spanish and Portuguese explorers and thence into China, the Philippines, and the west coast of Africa (Hedrick, 1919).

The guava is cultivated for its fruit in areas where oranges thrive and is badly damaged at temperatures below −2°C (28°F). Plants so injured often sprout from their roots. It will also slowly decline and eventually die if summer temperatures consistently fall below 16°C (60°F) (Wyman, 1986; Mowry et al., 1958). 'Supreme', 'Red Indian', and 'Ruby' are cultivars commonly grown in Florida. 'Kampuchea', 'Klom Toon', and 'Klom Sali' are popular in Malaysia, where guava-producing areas are located in Perak, Johore, Selangor, and Negri Sembilan.

Guava is propagated by seeding into light, sandy loam. This is especially true of *P. littorale* var. *longipes,* of which there are no named cultivars. The seeds do not retain their viability for any length of time and so must be planted immediately after harvest (Hartmann et al., 2002). Seedlings require a longer time to come into bearing than do vegetatively propagated plants. Cuttings are difficult to root, but there has been some success in using both hardwood and semihardwood cuttings treated with IBA and rooted under mist (Pennock and Maldonado, 1963). Shield and patch budding onto seedling stock also is quite successful (Jaffee, 1970). Air layering is a useful practice for the homeowner and one of the most important methods for commercial propagation of the common guava. Simple mound layering or girdling the stems with wire just below the point you wish roots to emerge, then treating the area with 500 ppm IBA in lanolin paste before mounding will also stimulate rooting (Majumdar et al., 1968).

Guava does best on well drained loam soils but has been grown on sandy tin tailing areas in Malaysia. Space the plants about 3 to 6 meters (10-20 feet) apart, depending upon region and species. Typical spacings for Florida are 3.6 × 4.5 meters (11.8 × 15 feet), for Australia 4 × 6 meters (13 × 20 feet) and for Hawaii 5.2 × 7.6 meters (17 × 25 feet). In Malaysia the newly planted guava are shaded immediately after planting and watered until established. Generally, the plant needs little subsequent supplemental irrigation except during fruit development. Little fertilizer is necessary and the plant does well with very little care other than normal weed control and corrective pruning. Plants are trained about four months after planting to the open heart, open center, or cup-shaped systems, and the first crop is produced two to three years after seedling planting or sometimes in about a year on budded stock. Fruit is mature about 90 to 150 days after bloom, though flowers and immature fruit can be found on the plant simultaneously. Flowers open between 5:00 and 7:00 a.m. and pollination is effected by bees. In Malaysia the young fruit is bagged about 45 days after bloom to protect them from pest attack and abrasion. In that country and in Hawaii, yields begin at about 10 metric tons per hectare (10 tons per acre) in the third year and increase to 30 metric tons (33 tons) after the tenth year. Production is much greater in Taiwan, where 90 metric tons per hectare (90 tons per acre) were not uncommon. Florida producers can expect to harvest about 400 pounds (181.4 kilograms) of fruit per tree.

The fruits contain about 152 milligrams of vitamin C per 100 grams of fruit, making them several times richer in this vitamin than oranges. They are processed into jellies, jams, juice, and sometimes are eaten fresh.

guelderberry: *See* **cranberry, highbush.**

guelder rose *(Viburnum opulus): See* **cranberry, highbush.**

haw, black: *See* **cranberry, highbush.**

haw, sweet *(Viburnum prunifolium): See* **viburnum, blackhaw.**

hawthorn: Rosaceae. The genus *Crataegus* contains less than 1,000 species of thorny shrubs and small trees native to the north temperate zone. Leaves alternate; flowers white or sometimes pink; fruit a small pome, usually red. Most species are hardy to zone 4 (Hortus, 1976). The fruit of many species has been used for preserves since ancient times. That of *C. coccinea* was eaten fresh or mixed with chokecherries and serviceberries, pressed into cakes, and dried for winter by western Indian tribes (Hedrick, 1919).

These plants will thrive in poor soils and are not finicky about soil pH. Plant them in full sun. Bushes may require maintenance pruning only, removing dead wood in early spring. Sow seeds in autumn for winter stratification. Seeds of some may require two years to germinate. Other methods of seed stratification are too involved for a full discussion here (Dirr, 1998). The fruit is used for pastries and jellies.

hazelnut: *See* **filbert.**

hill-gooseberry: Myrtaceae. DOWNY MYRTLE. *Rhodomyrtus tomentosa* Wight is an evergreen ornamental shrub 1.2 to 3 meters (4-10 feet) tall that thrives in deep, acid, sandy soils. The leaves are light green, smooth above, tomentose beneath, elliptic-obovate in outline, about 5 centimeters (2 inches) long and 2.5 centimeters (1 inch) wide. Flowers solitary or in twos or threes, rose-pink to light purple. The fruit is globose, and about 1.25 centimeters (0.5 inch) diameter, downy, greenish purple, and has soft, sweet purplish pulp. The fruit is used for jams. The plants are propagated by seeds (Mowry et al., 1958).

hobbleberry: *See* **cranberry, highbush.**

hobblebush *(Viburnum alnifolium): See* **cranberry, highbush.**

Holboella: Lardizabalaceae. There are about 10 species in the genus *Holboella.* All are monoecious, climbing shrubs native to the Himalayas and China. Leaves alternate; fruit a fleshy pod with many black seeds (Hortus, 1976).

Holboella coriaceae Diels. This shrub with purple fruit about 5 centimeters (2 inches) long is native to central China.

Holboella grandiflora Reaub. The purple fruit of this species, native to western China, reaches lengths of up to 15 centimeters (6 inches).

Holboella latifolia Wallch. GOPHLA, KOLE-POT. This native of the Himalayas produces purple, mealy, and insipid-flavored fruit up to 8 centimeters (3 inches) in length (Hedrick, 1919).

holly, mountain *(Prunus ilicifolia): See* **Prunus**.

hollygrape *(Mahonia aquifolium): See* **grape, Oregon**.

honeysuckle, Jamaica *(Passiflora laurifolia): See* **passionfruit**.

houndsberry *(Solanum melanocerasum): See* **huckleberry, garden**.

huckleberry: Ericaceae. This name is variously applied to members of the *Vaccinium* genus but is more properly reserved for *Gaylussacia* spp. This entry deals only with the latter. There are about 40 species of shrubs in the genus *Gaylussacia*. All are up to 2 meters (6.5 feet) tall and native to North America and South America. Leaves alternate, simple; flower white, pink, or red; fruit a 10-celled berrylike drupe distinguished from the fruit of *Vaccinium* by having 10 quite large, boney seed coverings. These are noticeable and a bit unpleasant when the fruit is eaten. The fruit is borne in small clusters and individual fruit may reach up to 6 millimeters (0.25 inch) in diameter. The fruit is mostly blue to black and sweet. There is no commercial production, but the fruit is harvested for local use.

Gaylussacia baccata (Wangenh.) C. Koch. BLACK HUCKLEBERRY. This species, hardy to zone 2, bears glossy black fruit and grows in eastern North America from the Canadian Maritimes south to Georgia and west to Iowa.

Gaylussacia brachycera (Michx). A. Gray. BOX HUCKLEBERRY. The fruit is blue and the evergreen plants grow in the mid-Atlantic states west to Tennessee. The plant grows very slowly and is hardy to zone 6 (Wyman, 1986).

Gaylussacia dumosa (Andr.) Torr & A. Gray. DWARF HUCKLE-BERRY. Plants are found along the Atlantic coast of the United States from Newfoundland to Georgia and are hardy to zone 2.

Gaylussacia frondosa (L.) Torr. & A. Gray. BLUE HUCKLEBERRY, BLUE-TANGLE, DANGLEBERRY, DWARF HUCKLEBERRY, TANGLE-BERRY. Plants bearing the dark blue, glaucous, acidulous fruit grow along the Atlantic coast of the United States from New Hampshire to Florida and are hardy to zone 5 (Hedrick, 1919; Wyman, 1986). The crop is harvested solely from the wild and the berries used in pies, puddings, jams, and jellies in a manner similar to blueberries. Pollination requirements for this genus are unknown, but pollination is likely carried out by insects, notably bees, as in *Vaccinium*.

To propagate the plants sow seeds of ripe fruit in autumn after harvest for winter stratification. The plants may also be divided in early spring before growth begins. Layering is effective, as is propagation of hardwood and softwood cuttings (Wyman, 1986). Plant the bushes about 1 meter (3 feet) apart in light to moderate shade in well drained acid soils high in organic matter. This is usually sandy soil overlain with a layer of peat or leaf mold (Hortus, 1976). The plants require little pruning except for maintenance and removal of dead or damaged wood.

huckleberry, blue *(Vaccinium membranaceum): See* **blueberry.**

huckleberry, box *(Gaylussacia brochycera): See* **huckleberry.**

huckleberry, California *(Vaccinium ovatum): See* **blueberry.**

huckleberry, evergreen *(Vaccinium ovatum): See* **blueberry.**

huckleberry, garden: Solanaceae. This plant is neither a true huckleberry nor a misnamed blueberry, but a relative of the tomato.

Solanum melanocerasum All. BLACK-BERRIED NIGHTSHADE, HOUNDSBERRY, PETTY MOREL, QUONDERBERRY, SOLANBERRY, SUNBERRY, WONDERBERRY. This is an annual plant growing to 0.7 meter (2.3 feet) in height. Its ripe, smooth, globular black fruit is about 1 to 2 centimeters (0.4-0.8 inch) in diameter, borne in clusters, and has the flavor of a bitter tomato. The fruit is cooked and eaten as is or made into pies and preserves. The leaves are sometimes used as potherbs. The plant is widely cultivated in western tropical Africa and a garden form is sometimes cultivated in home gardens, requiring about three months from seeding to harvest. Culture of this plant is similar to that of tomato (Hortus, 1976).

huckleberry, littleleaf *(Vaccinium scoparium): See* **blueberry.**

huckleberry, Montana blue *(Vaccinium membranaceum): See* **blueberry.**

huckleberry, shot *(Vaccinium ovatum): See* **blueberry.**

huckleberry, squaw: *See* **blueberry.**

huckleberry, sugar *(Vaccinium vacillans): See* **blueberry.**

huckleberry, thinleaf *(Vaccinium membranaceum): See* **blueberry.**

hullberry: *See* **blackberry.**

hurricane plant *(Monstera deliciosa): See* **monstera.**

I

imbe: Guttiferae. This shrub, *Garcinia livingsonei* Ander., is native to eastern Africa and is well adapted to both sandy and limey soils. The shrub reaches heights of 4.5 to 6 meters (15-20 feet); leaves stiff, leathery, oblong, dark green; flowers greenish-yellow and borne in the leaf clusters; fruit ovoid, 2.5 to 5 centimeters (1-2 inches) in diameter, orange, with pleasant flavored acidulous flesh. The plant is monoecious, with fruit maturing from April to August and again from October to January (Mowry et al., 1958). Propagation is entirely by seed.

Irish-mittens *(Opuntia vulgaris): See* **pear, prickly.**

islay *(Prunus ilicifolia): See* ***Prunus.***

ivry-leaves *(Gaultheria procumbens): See* **wintergreen.**

jamberberry *(Prunus fruticosa): See* **Prunus**.

jamberry *(Physalis ixocarpa): See* **cherry, ground**.

jaundiceberry *(Berberis vulgaris): See* **barberry**.

jostaberry *(Ribes nigrum* × gooseberry)*: See* **Ribes**.

jujube: Rhamnaceae. There are more than 40 species of spiny decid-uous or evergreen shrubs and trees in the genus *Zizyphus*. All are tropical or native to warmer regions of both hemispheres. Leaves al-ternate, simple; flower clusters small, green to light yellow, axillary; fruit a fleshy drupe, edible on some species. There are two species of horticultural interest, which furnish edible fruit.

Ziziphus jujuba Mill. CHINESE DATE, CHINESE JUJUBE, COMMON JUJUBE. Often a deciduous tree but sometimes a bush found from southeast Europe to China and hardy to zone 5b, though well hard-ened plants may withstand temperatures of –29°C (–20°F). The fruit is oblong to pyriform and about 3 centimeters (1.2 inches) in diameter with brownish skin and mild-flavored, crisp, sweet flesh, the sugar content of which often approaches 22 percent. If left on the tree, the fruit will dry as a fig. With protection it can be grown as far north as western New York where temperatures may reach –26°C (–15°F) but the fruit may not ripen fully in northern areas because of the abbrevi-ated growing season. This species has become moderately developed and many, mostly Chinese, cultivars have been introduced. The fruit of most cultivars requires two to four months from bloom to harvest.

Ziziphus mauritiana Lam. CHINESE DATE, COTTONY JUJUBE, DUNKS, INDIAN CHERRY, INDIAN JUJUBE, INDIAN PLUM. This ever-green species forms a large bush or sometimes a small tree. It is native to India but has been widely planted throughout warmer parts of the world, producing oblong orange to brown fruit up to 2.5 centimeters (1 inch) diameter. Seeds are surrounded by edible pulp. The plant is hardy to zone 10.

This species is less well developed than *Z. jujuba,* though some In-dian cultivars have been developed from improved selections. Fruit ripens about 180 days after bloom and is used candied, dried, or in preserves.

The species planted primarily in the United States is *Z. jujuba,* which is adapted to the hot, dry conditions of California and other areas in zone 9. Several cultivars bearing improved fruit up to 5 centimeters (2 inches) in length were developed in China, where the species has been cultivated for several thousand years (Hortus, 1976). China produces the fruit commercially, but there is no commercial production in the United States.

Ziziphus jujuba was long ago cultivated in the Mediterranean area for its edible fruit. From there it spread eastward into Persia and China. In southern India oil is extracted from the kernel of the fruit and in Syria the fruit is made into a substance resembling dry cheese (Hedrick, 1919).

The cultivars 'Li', 'So', 'Tanku', 'Vu', and 'Yu' are the best known. In the United States, 'Ling', 'Tusoa', and 'Li' are the most popular and all are self-fruitful (Hartmann et al., 2002; Wyman, 1986).

The plants can be grafted or T-budded onto jujube seedling stocks or propagated by hardwood cuttings or root cuttings, but seeds are the common method of propagation and will germinate best following a three-month warm/three-month cold cycle at 4°C (39°F) before planting (Hartmann et al., 2002; Dirr, 1998). Plants tolerate alkaline soils in the southwest United States and are adapted to hot, arid regions. The fruit is usually candied, dried, made into mincemeat, pickled, smoked, or canned, and the firm, just-ripe fruit can be eaten out of hand.

juneberry: *See* **serviceberry.**

juneberry, European *(Amelanchier ovalis): See* **serviceberry.**

juneberry, Florida *(Amelanchier florida): See* **serviceberry.**

juneberry, Korean *(Amelanchier asiatica): See* **serviceberry.**

juneberry, mountain *(Amelanchier alnifolia): See* **serviceberry.**

karanda: Apocynacae. This large, thorny shrub, *Carissa carandas* L., is native to India. Shiny leathery leaves are broadly ovate, about 5 centimeters (2 inches) long, with rounded or obtuse tips. Flowers fragrant, about 2 centimeters (0.75 inch) wide, white, and borne in small terminal clusters; fruit globose to ovoid, less than 2.5 centimeters (1 inch) long, purple-black when ripe, with pale red, acid pulp. This plant is not widely grown in North America. It thrives on several types of soil, requires little attention, and grows rankly when heavily fertilized. The species is propagated by seed, but seedlings grow slowly. The plants may also be propagated by air layering and cuttings. The fruit is too acidic to eat fresh but makes excellent jelly and drinks (Mowry et al., 1958).

kei-apple: Flacourtiaceae. UMKOKOLO. This dense, dioecious thorny shrub, *Dovyalis caffra* Warb., is native to South Africa, though well hardened shrubs can recover from temperatures of –7°C (20°F) to fruit the following season. It reaches heights of 5 to 6.7 meters (15-20 feet) under good growing conditions. Leaves are shiny green, about 5 centimeters (2 inches) long and are borne singly on new wood or in clusters on older wood. Flowers are yellowish, small, and inconspicuous. The fruit is rounded oblate, 2.5 to 3.8 centimeters (1-1.5 inches) in diameter, with thin yellow to greenish yellow, smooth skin and yellow, melting juicy flesh, ripening summer and autumn. The flesh is strongly acidic with the flavor of apricots. The plants are propagated by seeds, layers, and budding. For good fruit production space plants 3.3 to 4 meters (10-12 feet) apart. The fruit is used mainly for sauces, preserves, and jelly.

king's ace berry: *See* **blackberry.**

kinnikinic *(Arctostaphylos uva-ursi): See* **bearberry.**

kitembilla: Flacourtiaceae. CEYLON-GOOSEBERRY. This native of Sri Lanka, *Dovyalis hebecarpa* Warb., is somewhat more tender than the **kei-apple** but is still able to produce a crop in mid to northern Florida with adequate protection. The dioecious shrub reaches a height of about 5 meters (15 feet). The foliage is light green with leaves lanceolate to oval, 5 to 10 centimeters (2-4 inches) long and more or less velvety when young. The fruit is velvety, spherical, 2.5

centimeters (1 inch) in diameter, and maroon-purple. The flesh is purple, subacid to acid, juicy, and used for jellies and preserves.

kiwi: *See* **kiwifruit.**

kiwiberry *(Actinidia chinensis): See* **kiwifruit.**

kiwifruit: Actinidiaceae. BABY KIWI, CHINESE GOOSEBERRY, FUZZY KIWI, HARDY KIWI, KIWIFRUIT, STRAWBERRY PEACH, YANG-TAO. Several dozen species of viney, climbing shrubs comprise the genus *Actinidia.* All are native to Asia and have simple, alternate leaves, single or clustered flowers and bear berries. Most species grow wild in temperate regions of China.

Actinidia arguta (Siebold & Zucc.) Planch. ex Miq. BOWER ACTINIDIA, TARA VINE, YANG-TAO. This species is native to temperate areas of east Asia, Korea, northeastern China, eastern Siberia, and northern Japan and bears a small, grapelike fruit about 2.5 centimeters (1 inch) in diameter, weighing 5 to 8 grams (0.16-0.25 ounces). Many consumers prefer the flavor of this species' fruit to that of *A. chinensis.* The plant is hardy to zone 5 and because of this hardiness is beginning to find favor among small fruit growers in the northern United States and Canada where it has produced up to 23 metric tons of fruit per hectare (22 tons/acre) under optimal conditions. A well hardened plant will tolerate –32°C (–25°F).

Actinidia chinensis Planch. CHINESE GOOSEBERRY, KIWIBERRY, YANG-TAO. A native to Taiwan and China, this species is cultivated commercially in New Zealand for its fruit and is hardy to zone 8. Because of variability between hairy and smooth-skinned types, the kiwifruit of cultivation now is termed *Actinidia deliciosa* var. *deliciosa,* but sometimes still carries the older botanical name of *A. chinensis* (Ferguson, 1990).

Actinidia deliciosa var. *deliciosa.* KIWIFRUIT. This species, formerly *A. chinensis,* is the species of commerce.

Actinidia kolomikta (Rupr. & Maxim.) Maxim. The small grapelike fruit of this species is about 2 centimeters (0.8 inch) in diameter and of interest only on northern sites, where the plant can withstand winter temperatures of –35°C (–31°F). It bears very sweet fruit, the vitamin C content of which amounts to about 1 percent of its fresh weight. A native of temperate eastern Asia, this species is hardy to zone 5 (Hortus, 1976; Ferguson, 1990).

Actinidia polygama (Siebold & Zucc.) Maxim. SILVER VINE. The fruit of this species is salted and eaten in Japan but otherwise the plant is of little interest (Hortus, 1976).

All *Actinidias* grow well in either sun or partial shade. In 2004, Italy led in world production of this fruit with 365,000 metric tons (402,343 tons), followed by New Zealand with 320,000 metric tons (352,739 tons), and Chile with 130,000 metric tons (143,300 tons) (http://www.faostat.fao.org). The United States produced 24,000 metric tons (26,455 tons), almost totally in California which produced 25,900 tons (23,496 metric tons) (http://www.jan.mannlib .cornell.edu), with minor production in coastal South Carolina and Georgia. Oregon produces some hardy kiwi, and Canada has about 40 acres in production.

Actinidia has been gathered from the wild in China for thousands of years but cultivated for only about a century. Seeds and seedlings were introduced into Europe, the United Kingdom, the United States, and New Zealand early in the twentieth century and New Zealand began the first commercial cultivation of the fruit in the 1930s. Around 1970, growers in California, Italy, France, and Japan began to cultivate the fruit. The name "kiwifruit" was first used about 1960 with the initial exports of the fruit from New Zealand to the United States.

A. deliciosa var. *deliciosa* accounts for essentially all commercial production. The following discussion applies to this species except where otherwise noted.

'Hayward' originated in New Zealand from a seed brought from China in 1904 and has become the predominant cultivar because of its large fruit, fine flavor, and very good keeping qualities. The entire industry in New Zealand is now composed primarily of this one pistillate cultivar. Newer 'Hayward' strains are being developed. 'Abbott' and 'Monty' produce well in California. Staminate clones, such as 'Matua', with anthesis coincident with 'Hayward', are interplanted with the pistillate clone for pollination purposes. Cultivars of *A. arguta* that have performed well under Canadian trials include 'Geneva', 'Ananasnaya', 'Dumbarton Oaks', and 'Issai'. The male cultivars 'Meader' and '74-76' provide adequate pollination when planted as every third plant in every third row.

This plant is dioecious with functionally imperfect flowers, for although "pistillate" flowers appear to be perfect their stamens produce nonviable pollen. Staminate flowers have reduced ovaries. Neither flower type produces nectar. Small axillary inflorescences are borne near the base of flowering shoots. The following season new flowering axillary shoots are formed distal to the previous season's flowers, while the shoot apex remains vegetative. The plants bloom on current year's shoots produced from buds differentiated in the summer of the previous year. Although flower buds were induced the previous summer, differentiation occurs only in the following spring preceding bloom (Ferguson, 1990; Hopping, 1986). Pistillate flowers remain receptive to pollen for slightly more than a week after opening and require multiple bee visits for good fruit set. Staminate flowers shed pollen only for two to three days after opening. Because *Actinidia* flowers are not particularly attractive to bees, at least one hive per hectare (three hives per acre) is needed for adequate pollination. These are moved into the planting at 10 percent 'Hayward' flowering, that is, when about 10 percent of the pistillate ('Hayward') plants are in flower (Jay and Jay, 1984).

Kiwifruit can be propagated by cleft or whip and tongue grafting onto seedling rootstocks in late winter before the vines begin to bleed. Softwood cuttings taken from June to August and dipped in IBA prior to striking in a mist chamber root easily, as do semihardwood and root cuttings (Ferguson, 1990).

Actinidia does best on light, well drained soils high in organic matter and will not tolerate water-logged conditions. Soil pH should be slightly acid (pH 6-7) since the vines are subject to lime-induced chlorosis on alkaline soils. Adequate irrigation is necessary for good fruit production. Because of the vines' large leaf surface area, transpiration can exceed 95 liters (25 gallons) of water per day under some conditions. A frost-free growing season of about eight months is necessary for good fruit production. Avoid frost pockets and provide shelter from wind. Plants should be set in full sun but exposure of the vines and fruit to bright direct sunlight as a result of severe pruning or incomplete foliage cover can result in sunburned fruit and canes. Under ambient conditions of intense sun, as in California, light shading can prevent sunburn damage though there may be some reduction in flower bud induction in vines that are shaded in midsummer.

Vines are planted about 3 to 6 meters (10-20 feet) apart in rows spaced about 4.5 meters (15 feet) apart, depending upon the training system and intensity of cultivation. Traditionally, a ratio of one staminate vine to eight pistillate vines has been sufficient to provide adequate fertilization. Higher ratios of 1:6 or 1:5 can also be employed. Young plants are sensitive to excess fertilizer and salts, making the application of small amounts of fertilizer on a regular basis throughout the first growing season necessary. Precise needs for nutrients depend upon local conditions and the age and productivity of the vines. In general, apply about 165 kilograms nitrogen per hectare (150 pounds/acre) in two applications, with two-thirds applied in early spring and the last third after flowering. A phosphorus application of around 60 kilograms per hectare (54 pounds/acre) applied in late winter should be sufficient. Kiwifruit vines and fruit remove large amounts of potassium, which should be supplied at about 300 kilograms per hectare (268 pounds/acre) in split application in early, mid, and late spring. Always base fertilizer applications on a complete assessment of actual requirements (Smith et al., 1987).

Actinidia plants are vines and cannot support themselves. Allowed to trail on the ground they will produce a tangle of unproductive plants. They must therefore be supported. Two primary structures are in use for this support. The pergola provides for a single layer canopy of vines about 2 meters (6 feet) off the ground and can produce high yields. The T-bar system is less expensive to construct and maintain, though yields may not be as great as those in the pergola system. To construct the T-bar support, set sturdy posts 1 meter (3 feet) into the ground about 6 meters (20 feet) apart, with about 2 meters (6 feet) remaining above ground. Brace the end posts with dead men. A 5 × 10 centimeter (2 × 4 inch) crossbar about 2 meters (6 feet) long is fixed to and spans the posts in one direction, and three high strength 12 gauge wires are affixed to the cross-arms between posts. Canes are tied to the wires and their ends are allowed to overhang the outside wires (Beutel, 1986).

Place a stout stake beside each new plant and train a strong shoot along that stake to the center wire to form the trunk of the vine. Train the shoot in one direction along the wire to form a leader. The following spring train another shoot in the opposite direction along the central wire to form a second leader. Consider the leaders to be permanent. Remove all growth from below the graft union as it emerges. Train lateral shoots at right angles to the leaders and out along the wires. These become the bearing wood the following year. Tie all canes loosely in place and do not allow any shoot to twist around the stake or wire since this will weaken the wood. In early spring, prune pistillate plants by removing most of the wood that fruited previously and any twisted or damaged canes. Retain sufficient one-year-old wood along the wires to form a single layered canopy. Select replacement canes that form close to the leader and tie these along the wires. Laterals that previously fruited are cut back to about a half dozen buds distal to the most distal fruit. After two seasons remove and replace fruiting arms to reduce overcrowding. Pistillate plants may also require summer pruning in some areas. Prune staminate plants right after flowering by heading back the flowering arms to new growth near the leader.

Kiwifruit ripens from 180 to 200 days after bloom and is harvested as soon as possible, though those picked immature and allowed to ripen off the vine have poor color and flavor and a short shelf life. A delay in harvest can mean problems from weather and frost. In New Zealand, 6.2 percent soluble solids in the fruit juice is considered minimum for harvest. This is called the "maturity index." The minimum maturity index for California fruit is 6.5 percent. An index of 7 to 10 percent for both kiwifruit and *A. arguta* is necessary for highest quality following storage. Fruit of *A. arguta* ripen unevenly, making multiple harvests necessary with that species. Harvest the fruit by snapping it off its stalk. It will keep for several months in storage at 0°C (32°F) (Ferguson, 1990). The fruit is consumed fresh or processed into juice, jam, wine, and spirits. Unfortunately, processing destroys much of the unique flavor and bright color of the raw fruit. The stems of the plant are used for high quality paper or in the preparation of adhesives. The leaves are used for animal fodder, and essential oils are extracted from the flowers and seed (Ferguson, 1990).

kole-pot *(Holboella latifolia): See **Holboella.***

kronsbeere *(Vaccinium vitis-idea): See **lingonberry.***

lantern, Chinese: *See* **cherry, ground.**

lantern, Japanese *(Physalis alkekengi): See* **cherry, ground.**

lavacaberry: *See* **blackberry.**

laxtonberry: *See* **blackberry.**

lemon, water *(Passiflora laurifolia): See* **passionfruit.**

lingaro: Elaeagnaceae. This Philippine native, *Elaeagnus philippensis* Perr., is a climbing evergreen shrub related to Russian and autumn olives. The shrub reaches a height of about 3 meters (10 feet) and a diameter of 6 meters (20 feet) or more. Leaves small, oblong, pointed, light green above and silvery scurfy beneath; flowers borne in small clusters from the axils of the leaves on new growth; fruit pink or pale red and about the size of a small olive. The plant has performed well in southern Florida and is propagated by cuttings or by seeds, which require two to three weeks to germinate. Fruit flavor is tart but sweet and the fruit is used mainly for making a richly colored jelly. It may also be eaten out of hand if fully ripe (Mowry et al., 1958).

lingberry *(Vaccinium vitis-idea* L. *minus): See* **lingonberry.**

lingen *(Vaccinium vitis-idea* L. *minus): See* **lingonberry.**

lingenberry *(Vaccinium vitis-idea* L. *minus): See* **lingonberry.**

lingonberry: Ericaceae. This plant shares the genus *Vaccinium* with blueberry and cranberry and about 150 other species of deciduous or evergreen shrubs. All are native to the northern hemisphere and have simple, alternate leaves, white flowers borne either singly or in clusters, and a many-seeded berry.

Vaccinium vitis-idea L. AIRELLA ROUGE, ALPINE CRANBERRY, ARANDANO ENCARNADO, COWBERRY, CRANBERRY, DRY GROUND CRANBERRY, FOXBERRY, KRONSBEERE, LOWBUSH CRANBERRY, MOSS CRANBERRY, PARTRIDGEBERRY, RED WHORTLEBERRY, WHIMBERRY. Evergreen dwarf shrub up to 30 centimeters (12 inches) in height, with creeping rhizomes. The sour red fruit is about 1 centimeter (0.4 inch) in diameter and is borne on short uprights like the true cranberry. This species is found mostly wild in Europe and Asia and is hardy to zone 6.

Vaccinium vitis-idea L. var. *majus*. This varietas has leaves and fruit somewhat larger than the parent species. It blooms in spring and again in summer.

Vaccinium vitis-idea L. var. *minus* Lodd. MOUNTAIN CRANBERRY, ROCK CRANBERRY, LINGBERRY, LINGEN, LINGENBERRY, LINGON- BERRY. This plant is more commonly found growing wild in North America, is smaller than the species, and forms a dense mat of vege- tation. It produces bright red fruit somewhat smaller than that of the American cranberry. The plant will tolerate winters in zone 2 with ad- equate snow cover. The undersides of the leaves are cream colored with small dark dots, which distinguishes this plant from the bear- berry, the leaves of which are pale green and veiny beneath (St. Pi- erre, 1996). About 7.5 million kilograms (20 million pounds) are har- vested from the wild each year, mostly in Sweden and Finland, where the fruit is highly popular. Germany is the primary importer of this fruit. Annual Newfoundland harvests have approached 154,000 kilo- grams (401,890 pounds), with 38,000 kilograms (101,800 pounds) exported to Europe and the United States (St. Pierre, 1996). About 0.4 hectare (1 acre) is cultivated in Maine, another in Wisconsin, and about 6 hectares (15 acres) throughout North America. Germany cul- tivates about 20 hectares (50 acres) and perhaps 80 hectares (200 acres) are under cultivation worldwide (Markle et al., 1998). The fruit has been used as a substitute for true cranberry, which it resembles, and has been more popular in the Scandinavian countries than else- where. Aboriginal peoples consider it one of the most important ed- ible wild fruits in northern Canada (St. Pierre, 1996).

The lingonberry initiates flower buds in early autumn and bears its fruit in compact clusters at the ends of one-year-old shoots, called up- rights. A late frost during bloom, with temperatures of $-1.5°C$ ($29°F$), can result in 50 percent mortality of opened flowers. Temperatures of $-3°C$ ($27°F$) can cause 50 percent mortality of flower buds and unripe fruit. High temperatures above $25°C$ ($77°F$) and low temperatures be- low $10°C$ ($50°F$) inhibit pollen tube growth and result in poor fruit set (St. Pierre, 1996). The var. *majus* undergoes a second bloom in late summer or early fall in warmer areas and so can produce a good crop in this second fruit cycle if there is no early frost. The plants are partially self-fruitful but adequate cross-pollination can increase set by 30 to 60 percent. In the absence of adequate insect pollination, application of 500 ppm gibberellic acid at a concentration of 80 milliliters/square meter at 75 percent full bloom can induce parthenocarpic fruit set. 'Splendor' and 'Regal' (Wisconsin introductions), 'Sanna' and 'Susi' (Swedish introductions), 'Red Pearl' and 'Erntestegen' (German introductions), and 'Koralle', from the Netherlands, are the most popular cultivars. 'Sanna' has been very productive in Sweden, producing about 900 grams (29 ounces) of fruit per plant.

Seed germination is erratic and transplants from the wild are difficult to establish. The most successful method of propagation is to take shoot cuttings in late June or early July, strike them in peat plus 10 percent perlite, and place them in a mist bed with bottom heat. Hardwood or softwood cuttings taken in spring (April to June) and in autumn, dipped in 6,000 ppm IBA, and struck in peat in a mist bed show about 85 percent rooting in two to ten weeks (St. Pierre, 1996). Rhizome cuttings 5 centimeters (2 inches) in length and treated as shoot cuttings show about 70 percent rooting.

As with the other *Vaccinium,* lingonberry does best on a well drained soil high in organic matter and adjusted to pH 5. It tolerates upland soils but not poorly drained or extremely sandy soils and will show iron deficiency-induced chlorosis at near neutral soil pH. Low calcium/high organic matter (3 percent minimum) soils are best. There is some evidence that preplant incorporation of peat or sawdust into the soil at the rate of 3,000 kilograms/hectare (2,676 pounds/acre) increased plant growth, spread, and vigor (Stang et al., 1993). Soils should have at least 2 percent organic matter and be well drained, since compaction and waterlogging can increase the incidence of *Phytophthora* root rot. Maintain adequate moisture levels in the upper few inches of soil to promote vigorous rhizome growth. Space plants about 30 to 45 centimeters (12-18 inches) apart in rows spaced about 1 to 1.3 meters (3.25-4.25 feet) apart. As plants mature they form a solid bed of vegetation. An alternative is to space plants on 30 centimeter (12 inch) centers. Shading increases shoot growth but decreases yield; full sunlight provides the best conditions for growth and production.

Plants can show winter damage without snow cover. If snow cover is neither persistent nor sufficient, a winter mulch of straw will reduce the problem. Mulches of sawdust and peat 10 centimeters (4 inches) deep are best for weed suppression and also increase plant growth, yield, and fruit size, as well as the rate of plant spread. However, such mulches may promote earlier bloom and so increase the potential for frost damage.

This species has a low nutrient requirement and plants will suffer if overfertilized. In general, fertilizer applications recommended for bog cranberries will suffice for this species as well. Once plants are established in the first year, apply two light applications of a complete fertilizer. In later years, ammonium sulfate should be applied when growth first appears in spring and again four weeks later, at a rate of 28 kilograms per hectare (25 pounds/acre) per application. Usually, no pruning is required for the first six years. Subsequently, during the dormant season in late fall, alternate rows can be mowed to a height of 2.5 centimeters (1 inch) every three to four years for var. *minus,* and every two to three for var. *majus.*

Seedling plants require about three to six years to produce their first crop, while those that were produced vegetatively fruit the year after planting. Fruit ripen about 80 to 90 days after bloom (St. Pierre, 1996). Handpicking can yield 8 to 10 kilograms (18-22 pounds) per day of ripe berries. Harvesting with lowbush blueberry rakes can yield 40 to 45 kilograms (88-100 pounds) per day. The fruit varies from 0.5 to 1.2 centimeters (0.2-0.5 inch) diameter. Fruit will remain on the plant throughout winter if not harvested, but quality deteriorates rapidly. Yields vary from 500 to 1,000 kilograms/hectare (450-900 pounds/acre) in natural stands to 3,000 kilograms/hectare (2,676 pounds/acre) on fertilized forest sites and 5,000 to 9,000 kilograms/hectare (4,460-8,028 pounds/acre) on peat sites (St. Pierre, 1996).

Fruit is bitter when eaten raw but makes excellent processed products, such as jellies, preserves, sauces, pie fillings, bannock, and stuffing, and substitutes for the American cranberry. A frost improves the flavor of the fruit. The berries make an excellent drink when sweetened and diluted with water, ginger ale, or soda or when fermented into wine or liqueur. A mixture of lingonberry and rosehips makes an excellent jam (St. Pierre, 1996). The berries have been used as a digestive aid and supposedly stimulate gastric secretions. They also have been used to relieve indigestion and heartburn. The plants contain the compound arbutin that has been used to treat bladder and intestinal disorders and urinary tract infections. An ancient Breton legend has it that a young girl regrew her amputated arms when she bathed in an infusion of lingonberries. The stems and leaves mixed with alum produce yellow and red dyes. The fruit is rich in benzoic acid, which acts as a natural preservative and makes the fruit fairly acidic (pH 2.5). Fruit of var. *majus* contains more benzoic acid and is less susceptible to bacterial rots than that of var. *minus*. Fruit is also rich in vitamin C and pectin (St. Pierre, 1996).

loganberry: *Rubus ursinus* var. *loganobaccus* is a cultivar of trailing blackberry selected in 1883 by J.H. Logan of Santa Cruz, California, and hardy only to the Pacific coast states. Some researchers believe it is actually a blackberry × raspberry hybrid, most likely 'Texas Early' blackberry × 'Red Antwerp' red raspberry (Moore and Skirvin, 1990). A thornless mutant of 'Logan' was found in California in 1929. *See* **blackberry.**

love-in-a-mist *(Passiflora foetida): See* **passionfruit.**

lowberry: *See* **blackberry.**

M

malka *(Rubus chamaemorus): See* **blackberry.**

manzanita *(Arctostaphylus uva-ursi): See* **bearberry.**

manzanita, bigberry: *See* **bearberry.**

manzanita, Columbia: *See* **bearberry.**

manzanita, greenleaf: *See* **bearberry.**

manzanita, pine: *See* **bearberry.**

manzanita, woolly *(Arctostaphylus tomentosa): See* **bearberry.**

maypop *(Passiflora incarnata): See* **passionfruit.**

mealberry *(Arctostaphylus uva-ursi): See* **bearberry.**

mespilus, snowy: *See* **serviceberry.**

miltomate *(Physalis ixocarpa): See* **cherry, ground.**

miquelberry: *See* **wintergreen.**

missey-moosey *(Sorbus americana): See* **ash, mountain.**

molka *(Rubus chamaemorus): See* **blackberry.**

monox: *See* **crowberry.**

monstera *(Monstera deliciosa):* Araceae. BREADFRUIT VINE, CERI-
MAN, CUT-LEAF PHILODENDRON, FRUIT-SALAD PLANT, HURRICANE
PLANT, JAPANESE PINEAPPLE, MEXICAN BREADFRUIT, SPLIT-LEAF
PHILODENDRON, SWISS-CHEESE PLANT, WINDOW PLANT. The genus
Monstera contains about 25 species of mostly epiphytic climbers pri-
marily native to Mexico and Central America. U.S. production is cen-
tered in Hawaii and south Florida. Other production areas include
Mexico, Latin America, and tropical South America.

Monstera deliciosa Liebm. A large tropical vine climbing to 10
meters (33 feet) or more in its native habitat but usually less than 0.6
meter (2 feet) in height when grown as a pot plant. Leaves large, per-
forated, ovate and up to 1 meter (3 feet) wide; flowers up to 0.3 meter
(1 foot) long. The plant is primarily used as an ornamental (Hortus,
1976), but the fruit has a pleasant aroma and edible pulp and is eaten
mostly in the tropics. The multiple fruit resembles an ear of corn
about 20 centimeters in length which has a rind consisting of six-
sided plates that are the terminals of the berries. The plants are fairly
easily grown in rich loamy soil with good drainage (Wyman, 1986).
They are propagated by stem cuttings struck in sand or soil and kept
in a humid atmosphere, by air layering, or by seeds (Hartmann et al.,
2002).

The first harvest occurs about three to four years after planting.
Fruit is ripe when the platelets fall off. Fruit flavor resembles a mixture of banana and pineapple, but it is important to eat only the pulp of
the fruit; other parts of the plant may contain toxic amounts of oxalic
acid.

mooreberry *(Vaccinium uliginosum): See* **blueberry.**

mooseberry *(Viburnum alnifolium): See* **cranberry, highbush.**

moose bush *(Viburnum alnifolium): See* **cranberry, highbush.**

moosewood-hopple *(Viburnum alnifolium): See* **cranberry, high-bush.**

morel, petty *(Solanum melanocerasum): See* **huckleberry, garden.**

mortinia *(Vaccinium mortinia): See* **blueberry.**

nannyberry: *See* **cranberry, highbush** or **viburnum, blackhaw.**

natal-plum: *See* **carissa.**

nectarberry: H.G. Benedict discovered this fruit in California and introduced it in 1937. It may be a seedling of **youngberry** or a chimera of **boysenberry** and it is nearly identical to boysenberry in growth habit and production (Brooks and Olmo, 1972). *See* **blackberry.**

nessberry: An upright blackberry derived from a cross of *Rubus rubrisetus* (dewberry) × *R. stigosus* 'Brilliant' (red raspberry) by H. Ness of Texas and introduced in 1921. The fruit is larger than that of 'Logan' and has red skin and the flavor of raspberry but retains its core like the blackberry (Brooks and Olmo, 1972). *See* **blackberry.**

nightshade, blackberried *(Solanum melanocerasum): See* **huckleberry, garden.**

nut, conch *(Passiflora maliformis): See* **passionfruit.**

olallieberry: A hybrid of 'Black Logan' × youngberry. *See* **blackberry.**

oleaster *(Elaeagnus angustifolia): See* **olive, Russian.**

oleaster, wild *(Shepherdia argentea): See* **buffaloberry.**

olive, autumn: Elaeagnaceae. There are about 40 species of shrubs or small trees in the genus *Elaeagnus,* mostly native to southern Europe and Asia. Leaves are alternate; flowers are bisexual; fruit is drupelike and edible, though not palatable.

Elaeagnus umbellata Thunb. AUTUMN ELAEAGNUS. This deciduous shrub, native to the Himalayas, grows up to 6 meters (19.6 feet) in height and bears silvery-pink fruit following a very fragrant bloom. Under good environmental conditions the plant can escape and become a weed, as it has along the highways of southern New England and in areas of the southern United States. The plant is native to Japan and China, from whence it was introduced into the United States about 1830. It is hardy from zone 3 to 8 (Dirr, 1998).

The plant is fairly easily propagated from hardwood and root cuttings, layering, and grafting. Seedling propagation requires stratification, and germination sometimes takes two years (Hortus, 1976). The species does best on well drained, low fertility, loamy and sandy soils at slightly acid pH and has excellent drought tolerance. Plants require full sun for best growth and are usually used for conservation purposes. They are not the first choice for a landscape plant.

The sweet-tart fruit contains high levels of lycopene (15-54 milligrams/100 grams fruit). Tomato, a major dietary source of lycopene, contains only about 3 milligrams/100 grams fruit (Fordham et al., 2001). Lycopene is a carotenoid widely believed beneficial as protection against myocardial infarction and some forms of cancer (Kohlmeier et al., 1997; Clinton, 1998).

olive, Russian: OLEASTER, SILVER BERRY, TREBIZOND DATE, WILD OLIVE. This plant, *Elaeagnus angustifolia* L., forms a deciduous shrub or shrubby tree up to 6.5 meters (21 feet) in height. The yellow, axillary flowers are very fragrant. The small, silvery fruit is often described as drupe-like but more correctly is an achene covered by a fleshy perianth. The species is native to southern Europe and western Asia and hardy to zone 2. It does not tolerate conditions found in zone 8 and higher. The tree at one time was cultivated for its acid, mealy, but agreeable-tasting fruit, which was commonly sold in markets in Istanbul, Turkey. Formerly, the Iranians made a dessert from the fruit called "zinzeyd" and the fruit is used to make a spirit in other parts of the world (Hedrick, 1919). A sherbert is made from the fruit in the

Orient (Dirr, 1998). Plants can be propagated by seed treated with sulfuric acid for 30 to 60 minutes and fall seeded for stratification. There has been limited rooting success using 3,000 ppm IBA (Dirr and Heuser, 1987). *See* **olive, autumn.**

The species does best in light, sandy loam soils but will withstand dry and alkali soils, displaying a high degree of salt tolerance. Young plants are easily transplanted, but be sure to place them in full sun and to keep them pruned tightly. Plants of this species are susceptible to verticillium wilt and several other diseases and insects and may become invasive on good sites.

olive, wild *(Elaeagnus angustifolia): See* **olive, Russian.**

paniala *(Flacourtia cataphracta* Roxb.): This plant produces deep red to purplish, pleasantly flavored fruit about 2.5 centimeters (1 inch) in diameter. *See* **governor's plum.**

parson *(vitis labrusca): See* **grape.**

partridgeberry: *See* **lingonberry** or **wintergreen.**

passionflower: *See* **passionfruit.**

passionfruit: Passifloraceae. GRANADILLA, MAYPOP, PASSION-FLOWER, WATER LEMON. This family contains about 400 species of climbing vines mostly native to tropical America, though only about 40 produce edible fruit (Hartmann et al., 2002). Most are grown for ornamentals. The edible portion is the pulp covering the seeds, from which beverages are made. Leaves alternate, entire or lobed; flowers solitary or in clusters, bisexual; fruit a firm, many-seeded berry (Hortus, 1976). Most species are hardy to zone 10. In 1992, U.S. production was centered in Hawaii (6 hectares, 15 acres), Florida (15 hectares, 37 acres), Puerto Rico, and California (5 hectares, 12 acres) for a total of 959,000 kilograms (256,972 pounds). California produced about half this total poundage. Other production areas include South Africa, Australia, Kenya, and New Zealand (Markle et al., 1998).

Passiflora edulis Sims. GRANADILLA, PASSIONFRUIT, PURPLE GRANADILLA, PURPLE PASSIONFRUIT, YELLOW PASSIONFRUIT. Vines of this species are vigorous and may grow 4.5 to 6 meters (15-20 feet) annually with strong support. However, the plant usually lives no more than five to seven years. The egg-sized fruit is plum-colored to yellow when ripe. The tough outer skin encases many seeds, each in a juicy orange-colored pulp. The taste of the pulp is unique, musky, and guavalike, sometimes resembling that of an orange. The aroma and flavor of the yellow form are richer than those of the purple form. This species is cultivated for its fruit in Australia, New Zealand, Hawaii, Mexico, and southern California but is native to South America (Wyman, 1986).

The plant was introduced into Australia well before 1900 and from there seeds were brought to Hawaii in 1880. In Australia and South Africa hybrid cultivars under cultivation are grafted onto specific rootstocks. The main Hawaiian cultivar is 'Flavicarp'. The flowers are produced on new growth and plants are self-fruitful, with pollination best under humid conditions. Carpenter bees are the most efficient pollinators.

This species is usually seed-propagated. Extract seeds from the newly harvested fruit pulp and remove from the fruit and juice after a three-day fermentation period. Germination usually takes place about three weeks after planting. Cultivars may also be cleft-grafted onto seedling stocks. Scions of the purple-fruited hybrids are commonly grafted to seedling rootstocks of golden passionfruit *(P. edulis* f. *flavicarpa)* because of their resistance to *Fusarium* and nematodes (Doncaster, 1981; Teulon, 1971). The species is also sometimes propagated by layers and hardwood cuttings.

The plant is cultivated in the cool tropics where rainfall is at least 89 centimeters (35 inches) annually and well distributed. Generally, it will not tolerate a frost or high wind, and high temperatures reduce the fruit set. The best soils are sandy loams with neutral pH. Plants are fertilized with about 1.1 kilograms (3 pounds) of a 10 percent nitrogen fertilizer four times per year, trained to a trellis in a fashion similar to grape, and pruned annually either to stubs or canes.

In warm areas plants are pruned right after harvest, but in early spring in cooler regions. All weak shoots are removed and vigorous shoots headed back by a third of their length. The fruit turns quickly from green to purple or yellow when ripe and drops rapidly thereafter.

It will store two to three weeks at 10°C (50°F) and is said to be most tasty when slightly shriveled. Grafted vines produce fruit within one to three years of planting and about two to three months elapse between bloom and harvest.

The fruit is an excellent source of provitamin A and ascorbic acid and is used in beverages, desserts, jellies, and sometimes, fresh.

Passiflora foetida L. RUNNING POP, LOVE-IN-A-MIST, WILD WATER LEMON. This species is native to the American tropics but naturalized in the Old World tropics. The fruit is spherical, bright red to yellow, and edible.

Passiflora incarnata. MAYPOP.

Passiflora laurifolia L. BELLE APPLE, JAMAICA HONEYSUCKLE, POMME-DE-LIANE, VINEGAR PEAR, WATER LEMON, YELLOW GRANADILLA.

Passiflora ligularis Juss. SWEET GRANADILLA. The plant is native from Mexico to western Bolivia and is cultivated in the tropics for its sweet, yellow to purple fruit, which is generally considered superior to the fruit of *P. laurifolia.*

Passiflora maliformis L. CONCH APPLE, CONCH NUT, SWEET CALABASH, SWEET CUP. The fruit is dingy-colored when ripe but has a pale yellow, grape-flavored pulp at one time eaten with wine and sugar. It is native to the West Indies south to northern South America. Cultivation for its fruit is pretty much limited to the West Indies.

Passiflora mollissima. BANANA PASSIONFRUIT, CURUBA, SOFT-LEAF PASSIONFLOWER. This vine reaches heights of 3 meters (10 feet) and bears clear, rose-pink flowers. The plants prefer full sun but will tolerate partial shade. Prune them heavily after the second year of growth and provide a trellis for support.

Passiflora quadrangularis L. GIANT GRANADILLA, GRANADILLA. The greenish-yellow fruit is about the size of a small muskmelon and when ripe contains a translucent sweet-smelling pulp. The black seeds are eaten with the pulp and the plant is widely cultivated in the tropics.

pawpaw *(Asimina triloba): See* **banana, custard.**

peach, strawberry: *See* **kiwifruit.**

pear, cactus: *See* **pear, prickly.**

pear, Indian: *See* **serviceberry.**

pear, prickly: Cactaceae. CACTUS PEAR, PRICKLY PEAR CACTUS, SPINELESS CACTUS, TUNA. There are about 300 species of spiny shrubs or small trees in the genus *Opuntia,* many of which are commonly referred to as "prickly pear." The genus contains four subgenera, one of which has been further subdivided into five sections. Plants belonging to *Opuntia* subgenus *Opuntia* are the true "prickly pears" and "tuna." These grow in cool, semiarid climates at temperatures of 18 to 26°C (64-79°C). The genus is native to desert areas of northwestern Mexico and southwestern United States and was brought to Europe by early Spanish colonists from Mexico. It has been cultivated along the Mediterranean coast since the early 17th century (Retamal et al., 1987).

Opuntia ficus-indica (L.) Mill. INDIAN FIG, SPINELESS CACTUS. The plant is bushy, relatively void of spines, and grows up to 6 meters (20 feet) in height, with joints 0.3 to 0.45 meter (1-1.5 feet) in length. It produces yellow flowers up to 10 centimeters (4 inches) across. The edible fruit is mostly red but can be purple, white, or yellow and 5 to 9 centimeters (2-3.5 inches) long. It was highly popular at one time in California. This species is widely grown for its fruit and as a forage crop in the tropics and subtropics. Formerly the leaves were roasted by the Native Americans.

Opuntia humifusa (Raf.) Raf. This is a prostrate or spreading plant with fibrous roots native from Massachusetts to Montana and south to Florida and eastern Texas. The fruit is about 3 centimeters (1.2 inches) long and 1 centimeter (0.4 inch) wide, naked, and edible, with a sweet, somewhat acid taste. Native Americans roasted the leaves for food.

Opuntia lindheimeri Engelm. LINDHEIMER PRICKLY PEAR, TEXAS PRICKLY PEAR. This plant is an erect shrub to 4 meters (13 feet) in height with orbicular to obovate joints; spines one to six, pale yellow to white or brown and up to 3.5 centimeters (1.4 inches) long. The flowers are yellow to red and form purple fruit 3.5 to 7 centimeters (1.4-2.75 inches) long. The species is common mainly in Louisiana, New Mexico, and northern Mexico.

Opuntia phaeacantha Engelm. BASTARD FIG. This is a prostrate or sprawling cactus native from Texas to California and into northern Mexico. The fruit, with its sweet, juicy pulp, was relished by native American tribes and the leaves were roasted for food (Hedrick, 1919).

Opuntia vulgaris Mill. BARBARY FIG, BARBERRY FIG, IRISH-MIT-TENS, PRICKLY PEAR. A shrub or small tree up to 6 meters (20 feet) in height. The fruit is pear shaped, red, spineless, and up to 8 to 10 centimeters (3.2-4 inches) long. The plant was once widely planted as a host for the cochineal insects and is now naturalized in tropical and subtropical areas. This is a native to Central America but has long since been introduced into and cultivated in Europe. The salted leaves were sometimes used as food for seafarers. The cultivated fruit is very juicy and was once widely esteemed in Sicily where the plant flourishes on the lava slopes of volcanoes (Hedrick, 1919). Although the plant is usually grown in home gardens, there are some small commercial plantings, often on marginal land, in California and Texas, as well as Mexico and Chile. World production of fruit of the genus *Opuntia* is estimated at 304,800 metric tons (300,000 tons), of which Mexico produces 203,209 metric tons (200,000 tons) on 50,000 hectares (123,552 acres) (Flores and Gallego, 1994). Other important production regions include the Mediterranean basin, South America, South Africa, Australia, Mexico, Italy, and Chile. Cultivar development is local, with 'Reyna' and 'Esmeralbla' popular in Mexico, 'Morados' and 'Sanquina' in Spain, and 'Gialla' and 'Bianca' in Italy (Tous and Ferguson, 1996).

Sow seeds in sterilized soil or cleft graft plants. Plants can also be propagated by stem cuttings after allowing the cutting to suberize for a few weeks in a warm greenhouse before striking them. Use bottom heat to promote rooting, but high relative humidity is unnecessary (Carter, 1973; Edinger, 1970; VanDyk, 1983; Zieslin, 1980).

The plants do best on calcareous, open, well drained soils relatively low in organic matter and must never be held in wet soils. They require minimal care and usually will fruit within two to three years when started from stem cuttings. Four to eight months usually elapse between bloom and harvest. Spines located on the fruit rind are usually rubbed off by hand before harvest. The fruit peel readily separates from the pulp, which along with the seeds, are the edible parts of the plant. Harvest usually occurs from July to December, with a mature plant (eight years old) yielding 100 to 200 fruits (Knight, 1980). The fruit is used fresh, preserved, candied, or dried. The young cladophylls (leaves) are sometimes used as an ingredient of salsa or may be boiled and used like snap beans. They are also used for cattle fodder. The plant has been shown to possess some antidiabetic properties (Tous and Ferguson, 1996).

pear, swamp sugar *(Amelanchier oblongifolia)*: *See* **serviceberry.**

pear, sweet *(Amelanchier canadensis)*: *See* **serviceberry.**

pear, vinegar *(Passiflora laurifolia)*: *See* **passionfruit.**

pear, wild *(Amelanchier alnifolia)*: *See* **serviceberry.**

phenomenalberry: A second generation raspberry × blackberry hybrid, introduced by Luther Burbank in 1905. Also known as 'Burbank's Logan'. *See* **blackberry.**

philodendron, cutleaf *(Monstera deliciosa)*: *See* **monstera.**

philodendron, splitleaf *(Monstera deliciosa)*: *See* **monstera.**

pimbina *(Viburnum trilobum)*: *See* **cranberry, highbush.**

pineapple: Bromeliaceae. There are nine species of perennial herbaceous plants in the genus *Ananas.* All are native to tropical America and have narrow curving, spiny leaves forming a rosette. The simple inflorescence is borne on a scape and the fruit is a fleshy syncarp of fused ovaries. The plants are grown commercially and occasionally as ornamentals. There are two species bearing fruit generally considered edible.

Ananas bracteatus (Lindl.) Schult. RED PINEAPPLE, WILD PINEAPPLE. The red flowers give this plant its common name.

Ananas comosus (L.) Merrill. PINEAPPLE. This is the pineapple of commerce. The pineapple was once widely grown in Florida but its commercial production is now centered in Mexico, Puerto Rico, Hawaii, Taiwan, and the Philippine Islands. In 1995 Hawaii cultivated

840 hectares (20,800 acres). Guam produced 51 metric tons (50 tons), the Virgin Islands, 536 kilograms (1,437 pounds), and Puerto Rico, 59,700 metric tons (58,764 tons). Other production areas include Costa Rica, Dominican Republic, and Brazil. U.S. production continues to decrease, with 656,370 metric tons (646,000 tons) produced in 1986, 350,540 metric tons (345,000 tons) in 1995 and 270,000 metric tons (297,624 tons) in 2004 (Markle et al., 1998; http://www.faostat.fao.org). By 2004 Thailand led in world production with 1,900,000 metric tons (2,094,391 tons), followed by the Philippines with 1,759,290 metric tons (1,939,285 tons) and Brazil with 1,485,000 metric tons (1,581,816 tons).

The first Europeans to see the pineapple were voyagers with Columbus, who tasted the fruit at Guadeloupe Island in 1493. They were delighted and fascinated with the wonderful flavor. Records indicate that at least three kinds of pineapples were being grown in the New World by 1513. By 1549 the fruit was common in India. A monk in Mexico, in 1555, reported that the fruit was often preserved there in sugar. In 1595, Sir Walter Raleigh himself speaks of the abundance of pineapple grown in Guiana. One historian reported that the fruit was carried to China, probably from Peru, as early as 1592 and by 1599 had become naturalized in Java. It was commonly grown in China and the Philippines in the 17th century. The first successful planting in Europe was made in Holland in the 1680s and the first plants introduced into England came from there in 1690. Captain Cook planted pineapples on many Pacific islands in 1777. By that time they were also growing wild in Africa (Hedrick, 1919).

Many flowers are differentiated along the stem in a tight spiral just below the apex. Flower opening is acropetal, that is, from the base to the tip of the inflorescence. 'Smooth Cayenne' is grown in Hawaii and most other places. 'Red Spanish', until recently the most popular cultivar in the West Indies, is being replaced in that region by the larger 'Smooth Cayenne'. Because of its superior canning qualities 'Smooth Cayenne' makes up about 95 percent of the commercial production. 'Porto Rico' and the high quality 'Abbaka' sometimes are also grown. All pineapples are self-incompatible so that fruit in a field of a single cultivar will set no seed. If cross-pollination does occur, hardy bony seeds will set deep in the fruit flesh. Bloom can be induced by spraying the plants with 10 ppm naphthalene acetic acid

(NAA), often in spring, provided the number of leaves (70-80) is sufficient to assure development of the fruit.

The pineapple usually is propagated by suckers, by slips produced on the peduncle just below the fruit, or by planting the crown. Crown-formed plants are slow to come into bearing, needing about 24 months to produce a crop. Suckers are limited in number and require about 15 to 17 months. Both of these are rarely used in commercial production. Slips are the preferred method of propagation and are most abundant, requiring about 20 months to produce a crop (Collins, 1960).

The plants will tolerate a wide range of soils but do best in loose, light soils with good drainage. Plant in full sun or light shade. Avoid sites with strong winds since the top-heavy plants, with normally shallow root systems, are easily toppled. Allow slips or crown to dry for a few weeks before planting. Plant 0.3 to 0.6 meter (1-2 feet) apart in beds containing three to four rows. Set the slips about 10 centimeters (4 inches) deep, just enough to hold them upright. Slips planted in the fall rarely fruit until the second summer following planting, with cooler temperatures delaying the time of flowering and fruiting. Weeds are a great enemy of this shallow-rooted plant. Cultivation should be very shallow. In lieu of hand cultivation, the beds may be mulched with polyethylene. Fertilize as needed to keep the plant growing vigorously. As mentioned earlier, good fruit production depends upon 70 to 80 leaves having developed prior to the onset of fruiting. Fertilizer requirements vary considerably, but on sandy soil several applications of a complete fertilizer each year may be necessary, with supplemental nitrogen applications in winter and summer. Where excess lime and manganese are a problem iron may become limiting, calling for use of iron chelate supplements. Watch for other micronutrient deficiencies, particularly magnesium. The plants can withstand 0°C (32°F) but the fruit is damaged at temperatures of 5°C (42°F). If the plant is left in place after harvest it will flower each year, each time bearing more than a single fruit. This is called the ratoon crop. No more than two to three fruits should be allowed to mature in any single crop. However, with each succeeding ratoon crop the fruit and plants become smaller. Therefore, commercially, no fields are allowed to produce more than a couple of these crops. The fruit matures about six months after bloom, but ripening is not simultaneous. Fruit for shipping is harvested about a week before full maturity. Fruit used for fresh local consumption or which will be processed quickly is allowed to ripen on the plant. Cool summer temperatures may delay ripening enough so that the plants are exposed to damaging cold temperatures (Wyman, 1986). The pulp is eaten fresh, frozen, dried, or juiced. Offgrade fruit is used for crushed pineapple and the cores, trimmings, and juice from the crushing process are used for beverage juice. Good grade fruit is used for slices, chunks, and tidbits.

pineapple, Japanese *(Monstera deliciosa): See* **monstera.**

piprage *(Berberis vulgaris): See* **barberry.**

pitanga *(Eugenia pitanga): See* ***Eugenia.***

pitaya: Cactaceae. NIGHT-BLOOMING CEREUS, STRAWBERRY-PEAR. This Mexican cactus belongs to the genus *Hylocereus undataus* Britt. & Rose, and is often grown in warmer areas as an ornamental. It has the ability to climb trees and walls and has very large showy flowers, opening at night, and heavy, fleshy, three angled, jointed stems with short spines. The fruit is oval, red, and about 7.6 centimeters (3 inches) diameter, with soft, white, bland, juicy pulp. Plants are propagated by cuttings.

pitomba *(Eugenia luschnathiana): See* ***Eugenia.***

plum: *See* ***Prunus.***

plum, beach *(Prunus maritima): See* ***Prunus.***

plum, Indian *(Ziziphus mauritiana): See* **jujube.**

plum, nanny: *See* **cranberry, highbush.**

plum, sand: *See* **guava.**

poha: *See* **cherry, ground.**

pomegranate: Punicaceae. There are two species of deciduous shrubs or shrubby trees in the genus *Punica,* only one of which is of any import.

Punica granatum L. POMEGRANATE. This is a bushy shrub or small tree up to 6 meters (20 feet) high and widely cultivated in tropical and subtropical climates for both its fruit and ornamental value. The plant is native to southeast Europe and south Asia and has glossy leaves mostly opposite, single or clustered bisexual flowers borne near the ends of the branches, and fruit that is a leathery-skinned, purplish false berry up to 12 centimeters (4.7 inches) diameter with a persistent calyx and many reddish seeds surrounded by juicy pulp. The seed pulp is eaten, with soft-seeded cultivars having eating quality superior to that of the hard-seeded cultivars. The plant is hardy to zone 9.

Pomegranate is one of the oldest of cultivated fruit plants. U.S. production is centered in Hawaii and California, the latter having produced 12,800 metric tons (12,597 tons) of the fruit on 1,170 hectares (2,889 acres) in 1994. Annual world production was estimated at 812,800 metric tons (800,000 tons) (IBPGR, 1986), with major producing areas including Iran, India, Afghanistan, the Mediterranean countries, Mexico, China, Japan, and Russia (Markle et al., 1998).

This fruit has been cultivated since ancient times in Iran, Asia Minor, and the Himalayas, from whence it was distributed eastward into China. It was also known to the ancient Greeks and Egyptians at least 3,000 years ago and was brought to Rome from Carthage, where Pliny listed as many as nine different types of the fruit. It was introduced into the New World soon after Cortez conquered Mexico in 1521. Catholic missionaries had carried it to California by 1792, and it was observed growing in a semiwild state in Georgia as early as 1773. Because of its many seeds it represented to the ancient world procreation and abundance.

The fruit is borne terminally on short spurs produced on slow growing two- to three-year-old wood. Cultivars include 'Papershell', 'Spanish Ruby', and 'Wonderful', grown in California and Israel, 'Mollar' and 'Tendral' grown in Spain, and 'Schahvar' and 'Robab' in Iran. Pomegranate is both self- and cross-pollinated by insects.

The most successful manner of propagating this species has been through the use of 20 to 25 centimeter (8-10 inch) long, pencil thick hardwood cuttings taken in spring from one-year-old wood. These are struck 15 to 20 centimeters (6-8 inches) apart with the top 5 to 7.5 centimeters (2-3 inches) above ground (LaRue, 1980). Softwood cuttings and propagation by seeds or layering also are effective. Seeds need no after ripening but usually produce inferior plants. Since the hardwood cuttings root easily this plant is almost never grafted.

The plants do best on deep, heavy loam but are adapted to a wide range of well drained soils. The species thrives in hot, desert valleys. It can be grown throughout the tropics but produces good crops only under hot, semiarid conditions where there is sufficient heat during the growing season to ripen the fruit. This plant has a very low chilling requirement and will grow even in very warm areas. Dormant trees withstand temperatures as low as −12 to −9°C (10-15°F). Trees will fruit at three to four years of age under the best conditions and begin commercial production at ages five or six years. Plants are spaced about 5 to 7 meters (16.4-23 feet) apart, with closer spacings resulting in poor fruit color development and difficulty in harvesting. In general the bush requires little pruning, though plants will sucker badly and produce a dense, unproductive, spiny thicket if not thinned judiciously. Remove suckers, strong water sprouts, and excessive dense growth (Mowry et al., 1958). Otherwise train to multiple trunks to maintain a bushy form and ensure against fatal single-trunk rodent damage (LaRue, 1980). Light annual pruning will ensure new spur growth. Retain two-year-old or older wood for its fruitfulness.

Plants are fairly drought resistant but require normal watering to produce good fruit crops; overwatering results in soft, poorly-colored fruit. The pomegranate's fertilizer requirements are lower than those for other crops. Light applications of 0.2 to 0.4 kilogram (0.5-1 pound) of actual nitrogen with trace nutrients per tree per year is usually sufficient and need be applied only if growth yellows or if the plant shows poor shoot growth, since excess nitrogen lowers fruit quality.

Bloom to harvest time is four to ten months. Fruit is apt to split if left on the plant too long but will have poor quality if picked too early. It is normally harvested by clipping the stem just before full maturity. Fruit in good condition will store for several months in a cool, dry place. Average production of fruit in California is 2 to 2.4 metric tons per hectare (5-6 tons/acre). The fruit is used as fresh fruit or its juice expressed and used in cooking and jelly making. A balanced blend of the juice and sugar produces grenadine. The boiled rind has been used as a cure for tapeworm (Kumar, 1990). In addition to its fruit producing habit, the pomegranate is a highly ornamental plant often used in landscapes.

pomme-de-liane *(Passiflora laurifolia): See* **passionfruit.**

pop, running *(Passiflora foetida): See* **passionfruit**.

Prunus: Rosaceae. This is a large and diverse genus containing more than 400 species of trees and shrubs, including peaches, apricots, plums, and cherries. All are mostly native to the northern hemisphere and have alternate, simple, usually serrate leaves, bisexual flowers, and drupe fruit. When spoken of collectively and in reference to their fruit, the plants are usually grouped under the headings "stone fruit" or "drupe fruit." This entry will discuss only the shrub-type *Prunus* within the subgenus *Prunocerasus,* and will include the so-called bush plums, bush cherries, bush apricots, and the cherry-plum and plum hybrids. The true plums, unlike the cherries, have a waxy bloom over the fruit and lack the pubescence and furrowed stones of the apricots. Plum species produce single axillary buds rather than the cluster of three buds common in the almonds and peaches. The plums form a strongly hybridized group for commercial production, although production of the "bush" type plums is usually relegated to single-species types. Because of their diversity, plums ripen fruit anywhere from 70 to 120 days after bloom.

Prunus americana Marsh. AMERICAN PLUM, AMERICAN RED PLUM, AUGUST PLUM, BLACKTHORN, GOOSE PLUM, HOG PLUM, NA- TIVE PLUM, RIVER PLUM, SLOE, WILD PLUM. This species forms an often thorny, shaggy-barked shrub or shrubby tree up to 8 meters (26 feet) in height, growing wild from New England to Manitoba and south to Florida and New Mexico. It is hardy to zone 4. Cultivars include 'Black Hawk', 'DeSoto', 'Wolf', and 'Hawkeye', which produce red or yellow fruit about 2.5 centimeters (1 inch) in diameter which contains a flattened stone.

Prunus armeniaca var. *mandshurica* Maxim. MANCHURIAN BUSH APRICOT. A native to Korea and Manchuria, this plant is hardy to zone 6.

Prunus armeniaca var. *siberica* C. Koch. SIBERIAN BUSH APRI-COT. A native to East Asia, this plant is hardy to zone 5, but bears very small fruit. The cultivars 'Mandan', 'Scout', and 'Manchu' belong to this varietas.

Prunus besseyi L.H. Bailey. BUSH CHERRY, SAND CHERRY, WEST-ERN SAND CHERRY. Found from Manitoba to Wyoming and Colorado and east to Minnesota, this plant produces purplish-black, sweet fruit about 1 centimeter (0.4 inch) in diameter, the stones of which are rounded at both ends. This species is sometimes used as a rootstock for other *Prunus,* but suckers badly and is shallow-rooted (Baird, 1950). It is propagated by budding onto native plum seedlings or sometimes onto *P. besseyi* seedlings. The species is hardy to zone 3, but there are a few selected cultivars available.

Prunus cerasifera J.F. Ehrh. CHERRY PLUM, JAPANESE CHERRY PLUM, MYROBALAN PLUM, PURPLE-LEAF PLUM. This plant forms a deciduous shrub or small shrubby tree up to 8 meters (26 feet) in height. The leaves are red-purple with bright red tips. The pink flow-ers give way to spherical fruit about 2.5 centimeters (1 inch) diameter, sweet, yellow or red, with slightly flattened stone. Fruiting is spo-radic. The plants are native from Asia to the Balkans but have become naturalized from cultivation in Europe and North America. The cultivars 'Thundercloud' and 'All Red' are hardy to zone 4. The fruit, when it is produced at all, is used fresh and in jellies and preserves. The plant is sometimes used as a rootstock for other *Prunus* spp., par-ticularly for European and Japanese plums.

Prunus ×*cistena* N.E. Hansen. PURPLE-LEAF SAND CHERRY. This weak shrub usually is grown as an ornamental, although its black to purple fruit, which is rarely produced, is edible. The plant is hardy to zone 3.

Prunus depressa Pursh. BUSH CHERRY, SAND CHERRY. This plant is closely related to *P. pumila* and has trailing branches that form mats up to 2 meters (6.5 feet) in diameter. The dark-colored fruit about 1 centimeter (0.4 inch) in diameter is acid but fairly pleasant tasting. The stone is pointed at both ends. The plants are found from Quebec to Ontario and south to Pennsylvania.

Prunus fruticosa Pall. EUROPEAN DWARF CHERRY, EUROPEAN GROUND CHERRY, GROUND CHERRY, JAMBERBERRY, SIBERIAN DWARF CHERRY. This is a bush up to 3 meters (10 feet) in height that bears sour, red to purple, globe-shaped fruit about 0.5 centimeter (0.2 inch) diameter and which resembles small specimens of 'Early Richmond'. It is native to Europe and Siberia and hardy to zone 4, growing farther north than other sour cherries (Hansen, 1940).

Prunus gracilis Engelm. & A. Gray. OKLAHOMA PLUM, PRAIRIE CHERRY. The species forms a large bush up to about 5 meters (16 feet) in height that is found in the wild from Arkansas to Texas, being hardy to zone 6. The reddish fruit is covered with a light bloom and is about 1 centimeter (0.4 inch) diameter.

Prunus humilis Bunge. A shrub growing to about 2 meters (6.5 feet) in height, native to China, and hardy to zone 6, this species produces red fruit about 1 centimeter (0.4 inch) in diameter.

Prunus ilicifolia (Nutt.) Walp. EVERGREEN CHERRY, HOLLY-LEAVED CHERRY, ISLAY, MOUNTAIN HOLLY, WILD CHERRY. This species produces an evergreen shrub or shrubby tree from 1 to 8 meters (3-26 feet) in height and is found from California to northern Baja California. It bears red or rarely yellow egg-shaped fruit about 1 centimeter (0.4 inch) long having a thin, sweet pulp and smooth stone.

Prunus incana (Pall.) Batsch. An erect shrub up to 2 meters (6.5 feet) in height that produces very small, round, red fruit about 0.5 centimeter (0.2 inch) in diameter. It is native to Southeastern Europe and Asia Minor and hardy to zone 6.

Prunus incisa Thunb. CHERRY. A large shrub or shrubby tree producing purple black fruit about 1 centimeter (0.4 inch) in diameter. It is native to Japan and hardy to zone 6.

Prunus japonica Thunb. BUSH CHERRY, FLOWERING ALMOND, JAPANESE BUSH CHERRY, JAPANESE PLUM. This species forms a small shrub up to 1.5 meters (5 feet) in height that produces a more or less round, wine-red colored fruit about 1 centimeter (0.4 inch) in diameter. Its stone is pointed at both ends. It is native to China and Korea, hardy to zone 4, and much cultivated in Japan as an ornamental and for the fruit. The cultivars 'Alba', 'Kuliensis', 'Koziensis', 'Kinkiensis', 'Rosea', and 'Rubra' are available.

Prunus maritima Marsh. BEACH PLUM, SHORE PLUM. This species forms a straggly bush up to 3 meters (10 feet) in height, native from

Maine to Delaware along the coastal areas. The purple to yellow fruit, about 1 to 2 centimeters (0.4-0.75 inch) in diameter, is used locally for jams, sauces, and jellies. The plant is used to hold soil and sand along the coastal dunes and is propagated by seeds or root cuttings harvested and planted in the autumn. The cultivars 'Autumn', 'Raritan', and 'Stearns' are available.

Prunus nigra Ait. CANADA PLUM, MANITOBA NATIVE PLUM. This plant closely resembling *P. americana* is found growing wild from Quebec to Manitoba and south to Georgia and Louisiana and is hardy to zone 2. Cultivars include 'Assiniboin', 'Cheney', 'Hasca', and 'Oxford'.

Prunus pumila L. BUSH CHERRY, DWARF CHERRY, SAND CHERRY. A decumbent shrub growing up to 9 meters (30 feet) in height and found on the sandy shores along the Great Lakes. It is hardy to zone 4 and bears a purple black fruit about 1 centimeter (0.4 inch) diameter. The fruit is quite astringent, with a subglobose stone. This species is sometimes used as a rootstock for the native plums.

Prunus reverchonii Sarg. HOG PLUM. A shrub up to 2 meters (6.5 feet) in height that forms thickets naturally. Native to Oklahoma and Texas and hardy to zone 6, it produces round mostly yellow fruit with red streaks and an elliptic, reticulate stone. The fruit is about 1 to 2 centimeters (0.4-0.75 inch) diameter.

Prunus spinosa L. BLACKTHORN, SLOE. A rigid, deciduous suckering shrub or small tree from 1 to 4 meters (3-13 feet) in height. The fruit is globe shaped, about 1 centimeter (0.4 inch) diameter, blue to black, and has an astringent flavor. The plant is native to Europe and western Asia where the wood is used for turning stock and the fruits to flavor liqueurs. The species is hardy to zone 5.

Prunus tomentosa Thunb. BUSH CHERRY, CHINESE BUSH CHERRY, CHINESE BUSH FRUIT, DOWNY CHERRY, HANSEN'S BUSH CHERRY, MANCHU CHERRY, MONGOLIAN CHERRY, NANKING CHERRY. This is a compact shrub or small tree up to 2 meters (6.5 feet) in height that flowers profusely in very early spring and bears nearly stemless, scarlet fruit about 1 centimeter (0.4 inch) diameter. Cultivars are self-unfruitful and not as productive as other bush cherries. This species crosses readily with apricot, plum, and other cherries. It is native to temperate zones of eastern Asia and hardy to zone 3. The fruit is produced locally in the northern United States and also in Canada and

Russia. There is no commercial production. Plants of this species are propagated by budding onto Nanking cherry stocks, by seeds, or by suckers. The plant is adapted to lighter soils and will not tolerate heavy, compacted, or waterlogged soils (Yeager, 1935; Hansen, 1990; Howard and Brown, 1962). The cultivars 'Drilea', 'Eileen', and 'Baton Rouge' have been selected. The tangy fruit, which resembles that of the tart cherry, is used for pies, jams, and jellies. However, the fruit is extremely soft, limiting its shipping ability, and has a short shelf life. The plant is often used as a hardy ornamental.

Prunus virginiana L. BLACK CHOKECHERRY, CABINET CHERRY, CHOKECHERRY, CHUCKLEY-PLUM, RUM CHOKECHERRY, SLOETREE, WILD BLACK CHERRY, WILD CHERRY, WHISKEY CHOKECHERRY. This forms a shrub or small tree and may form thickets in the wild. The plants are very competitive and have a productive lifespan of about 40 years (St. Pierre, 1993). The small red fruit is a true cherry about 4 to 9 millimeters (0.15-0.35 inch) in diameter, borne in clusters of about two dozen fruit, and becomes purple to black, acid and quite astringent when ripe. The plant is native from Newfoundland to Saskatchewan and south to North Carolina and Kansas. It is hardy to zone 2 and was introduced into cultivation in 1724. This species is highly variable, with several forma and varietae documented.

Prunus virginiana var. *demissa* (Nutt.) Sarg., found in British Columbia, Alberta, Washington, Idaho, and California, forms dark red fruit. Plants of this varietas may reach heights of 6 meters (20 feet) and have heart-shaped leaves covered with very fine hair. *Prunus virginiana* var. *melanocarpa* (A. Nels.) Sarg. produces black fruit from California to the Rocky Mountains and north into British Columbia, Saskatchewan, Alberta, and Manitoba. It also reaches heights of about 6 meters (20 feet). This varietas has two forms, f. *melanocarpa* (black fruit) and f. *xanthocarpa* (yellow fruit). The f. *xanthocarpa* produces fruit slightly less astringent than the species type. *Prunus virginiana* var. *virginiana* produces a large shrub that reaches heights of 15 meters (50 feet) on good sites and is widely grown across Canada and the United States. Its f. *virginiana* produces crimson to deep red fruit, while its f. *leucocarpa* produces white to yellow fruit. The fruit was widely used by Native Americans dried and ground into a powder for use in soups, stews, and pemmican. The boiled bark was a remedy for diarrhea, and teas made from boiled twigs were used to relieve fevers and coughs and as a sedative. Dried roots were chewed and placed upon a wound to stop bleeding and the fruit was used to treat canker sores and ulcers (St. Pierre, 1993).

The chokecherry is noted for its toxic properties. Children and livestock have been poisoned by ingestion of leaves, seeds, and stems. The inner bark, buds, flowers, seeds, and suckers of this plant contain the cyanogenic glycoside prunasin which, when digested by stomach enzymes, becomes hydrocyanic acid (prussic acid) and results in cyanide poisoning. The plant tissue maintains its highest toxicity in spring and summer and the leaves become nontoxic when the fruit matures. Because of their greater concentration of hydrocyanic acid (243 milligrams/100 grams), wilted leaves pose more of a poisoning hazard than fresh leaves, which contain about 143 milligrams/100 grams. The concentration in wilted leaves is about 10 times the level at which poisoning symptoms can occur. Poisoning occurs when an animal consumes 0.25 percent of its weight in leaves over a 30 to 60 minute period (St. Pierre, 1993). This is equivalent to about 0.7 kilograms (1.5 pounds) of leaves for cattle and 0.1 kilogram (0.25 pound) for sheep.

Flower buds are formed on current year's wood and bloom on one-year-old wood. Since the plants flower later in the spring they are not especially subject to spring frost damage. The plant is generally self-fruitful and insect pollinated, with daytime temperatures below 30°C (86°F) and cool nights favoring fruit set. The use of a different cultivar for every tenth plant in the row, and supplemental honeybee introductions, will improve fruit set. Plants may begin fruiting in their second year but more usually three to four years are required for first fruit.

Plants do best in full sun and moisture-retentive but well drained soils and are intolerant of salty soils. Soil pH is not critical, with the optimum reported as being between 6 and 8. The cultivar 'Boughen Sweet', introduced from Manitoba prior to 1923, produces large, mild fruit. 'Canada Red' produces leaves that turn a deep purple by autumn and high yields of large, black fruit with excellent flavor. 'Johnson', introduced from Minnesota, produces high yields of good quality fruit and the tree has high ornamental value. 'Mission Red' and 'Mission Yellow' both are excellent ornamentals and produce colorful fruit that is excellent for wine. 'Schubert', introduced from

Bismarck, North Dakota, produces plants that have a pyramidal habit with dense, purple foliage tinged with green. It produces good yields of large black fruit that is excellent for jams, jellies, and wines.

This species may be propagated by seed, suckers, rhizome cuttings, semihardwood cuttings, crown division, or grafting onto *P. padus* (nonsuckering) stocks, and through micropropagation. Seedling plants are variable and often do not resemble the parent type. Sow seed outdoors in the autumn about 1 centimeter (0.4 inch) deep. Germination rates vary from 30 to 70 percent and only about one fourth of the seedlings will be usable. Seedlings require about one to two years to reach transplant size.

Transplant plants when about 0.6 to 1 meter (2-3 feet) in height in the early spring or late fall when dormant, spacing them about 2 meters (6.5 feet) apart in rows about 6 meters (20 feet) apart (St. Pierre, 1993). A dry falls followed by a cold winter can substantially increase mortality of fall-planted stock. Cut back spring-set plants to about 20 centimeters (8 inches) of the ground to promote branching (Patterson, 1957; St. Pierre, 1993). Cut back fall-set plants in the following spring. Irrigate with water having a salinity of less than 1 mS/cm (millisiemens/centimeter). In arid areas newly established plants require about 4.5 liters (1 gallon) of water per week. By the second year plants will need about 18 to 23 liters (4-5 gallons) every 2 weeks. Excessive irrigation will contribute to cracked and insipid fruit, root rot, and poor mineral absorption (St. Pierre, 1993).

Fertilizer requirements have not been determined for this species. However, the following general recommendations should prove adequate. Nutrient levels on prairie soils should be maintained at about 35 kilograms (77 pounds) nitrogen, 27 kilograms (60 pounds) phosphorus, and 136 kilograms (300 pounds) potassium per 0.4 hectare (1 acre). Short terminal growth and pale green color of leaves suggests the need for increased fertilizer application. While plant growth is young and active apply about 45 kilograms nitrogen, 6 kilograms phosphorus, and 56 kilograms potassium per hectare per year (40 pounds nitrogen, 5 pounds phosphorus, and 50 pounds potassium per 1 acre). For mature plantings, apply about 17 kilograms nitrogen, 3 kilograms phosphorus, and 34 kilograms potassium per hectare per year (15 pounds nitrogen, 3 pounds phosphorus, and 30 pounds potassium per acre) based upon assumed yields of 5,600 kilograms per hectare (5,000 pounds per acre). Apply fertilizer in early spring or late fall.

Prune plants in early spring. Remove weak and damaged wood, low, spreading branches, and other wood to keep the center of the shrub open to sunlight and good air circulation. Remove older, nonproductive wood so that most of the shoots are one to four years old. These are most productive. Prune out cankers and other diseased tissue immediately upon notice and disinfect pruning shears with rubbing alcohol. Prune leaders to a height of 2 to 3 meters (6.5-10 feet) to maintain proper plant height.

The fruit long have been favored for jellies, syrups, sauces, jams, and wine in the prairie states and provinces of the United States and Canada. Fruit matures about ten weeks after bloom. Entire clusters may be harvested and the fruit stripped from the cluster prior to use. Alternatively, individual fruit may be harvested mechanically or by using a berry rake or vibrator. Each shrub will yield about 13.6 kilograms (30 pounds) of fruit, with 770 to 2,470 berries per kilogram (350-1,100/pound). Expect yields of about 11,200 kilograms per hectare (10,000 pounds/acre) (St. Pierre, 1993).

Three loosely associated groups of shrubby or semitree form plums and plum hybrids stretch across species lines:

> *Plum group 1:* The native plum group comprises species native to North America, particularly *P. americana* and *P. nigra,* which are noted for their coldhardiness, although the fruit is only of fair quality. This group includes the cultivars 'Assiniboin', 'Bounty' and 'Cheney' *(P. nigra),* and 'Chilcott', 'Manet', 'North Dakota', and 'DeSoto' *(P. americana).* Cultivars in this group produce abundant pollen and are the second to bloom in spring, following the hybrid plums. They often are used to pollinate hybrid cultivars at a ratio of one plant of this group to six plants of the hybrids in groups 2 and 3.
>
> The American plums are often grafted to wild plum or Myrobalan stocks or can be propagated from autumn-sown seed. Plants do best on clay loams but will tolerate other soil types so long as they are well drained (Blair, 1954). Space plants about 3 meters (10 feet) apart in rows spaced about 7 meters (23 feet) apart on sites with good air drainage. This will re-

duce the chances of damage by spring frosts. The plants usually bear on spurs and plants are pruned by lightly thinning the outside limbs each spring to give sunlight to spurs and so invigorate them. Head the plants back only lightly (Blair, 1954). Light applications of about 0.4 kilogram (1 pound) of ammonium sulfate per plant applied in early spring is usually sufficient.

Plum group 2: This is a group of hybrids between the native plums and the Japanese plum *(P. salicina),* or between the native plums and some other species, such as *P. simonii,* the Chinese apricot plum. They have mostly replaced native species in cultivation although they are generally not as hardy or prolific. Many plants in this group are not inter-fruitful, are poor pollen producers and are therefore often interplanted with native plums for cross-fertilization. The native × Chinese apricot plums will cross-pollinate the other hybrid plums. Cultivars include 'Pipestone', 'Pembina', 'Cree', and 'Fiebing' (native × *P. salicina*), 'Grenville' (Burbank × native plum), and 'Tokata', 'Kaga', 'Superior', and 'Hanska' (native × *P. simonii*). The fruit of cultivars of native × *P. simonii* origin have a distinctly pleasant, apricot-scented flesh.

Site selection and requirement are similar to those for other plum types. Set one- or two-year-old bushes about 5 meters (16.5 feet) apart and train them to bush form. These plants need more pruning than those of the American plums since they bear on one-year-old wood. At planting head to about 0.3 meter (1 foot) above the ground. In the second year, prune to five to six strong branches and head them back by about one fourth their length to promote vigorous shoot growth. Fertilizer requirements are similar to those of American plums. The fruit must be harvested when well colored but still firm, since overmature fruit drops quickly.

Plum group 3: The third group is composed of hybrids between the native plums and *P. besseyi,* or between the Japanese plums and *P. besseyi.* This group also is sometimes called the cherry-plum, or plum-cherry, hybrids. The plants are very hardy, bloom the latest of the three groups, come into bearing early, hold and ripen their fruit early, and are drought resistant. They have better flavor than the native plums. However, they are generally short-lived (eight to ten years), poor pollen producers, self-sterile, and bear small fruit with poor keeping quality. They are intolerant of air pollution. The hardiest cultivars include 'Oka', 'Opata', and 'Compass', followed by 'Sapa'. This cultivar often bears on one-year-old wood and therefore winterkill becomes a problem. Sunscald also may be a problem in some areas (Baird, 1954; Yeager, 1935).

Site and soil requirements are similar to those of other plums. Set the plants about 2 to 4 meters (6.5-13 feet) apart and cut them back to 10 centimeters (4 inches) from the base of the branches at planting to induce branching near the ground. Although they are hybrids of shrubs and shrubby trees, they are best grown as shrubs since pruning them to a tree form would destroy too much of the bearing wood. Plants of this group bear on one-year-old wood (Hansen, 1990). Head back the branches in the second spring to encourage lateral branching; little further pruning is necessary. When older plants begin to lose their vigor, cut them back severely by half. You may wish to cut back half of the plant in one year and half the next to avoid "weedy" growth and overproduction (Patterson, 1957). Fertilize plants in this group sufficiently to promote vigorous new shoot growth.

pruterberry *(Ribes nigrum × R. divaricatum × R. uva-crispa): See* ***Ribes.***

Quebec berry *(Amelanchier stolonifera): See* **serviceberry.**

quince, Chinese *(Chaenomeles cathayensis): See* **quince, flowering.**

quince, flowering: Rosaceae. The genus *Chaenomeles* contains three species of mostly deciduous shrubs native to east Asia. The plants have alternate leaves and single or clustered waxy flowers that appear in early spring before the leaves. The fruit is a hard, quince-like pome. The plants are mostly hardy to the north and the fruit of some species is used in preserves and jellies.

Chaenomeles cathayensis (Hemsl.) C.K. Schneid. CHINESE QUINCE. This plant can reach heights of up to 3 meters (10 feet), producing a spiny shrub with white to pink flowers and bearing abundant fruit.

Chaenomeles japonica (Thunb.) Lindl. ex Spach. LESSER FLOWERING QUINCE. This is a spiny dwarf shrub up to 1 meter (3 feet) high that bears fruit resembling a small, gnarled apple. The plant is native to Japan, hardy to zone 4, and was introduced into Europe in 1815. In the United States the fruit was used in jellies and was sometimes eaten baked or stewed.

Chaenomeles japonica var. *maulei*. MAULES QUINCE. A variety of the lesser flowering quince that produces single, salmon to orange flowers.

Chaenomeles speciosa (Sweet) Nakai. JAPANESE QUINCE, FLOWERING QUINCE. A spiny shrub somewhat larger than *C. japonica,* reaching heights of 2 to 3 meters (6.5-10 feet). The plant is native to China and hardy to zone 4. Fruit is a fragrant, speckled pome 5 centimeters (2 inches) in length, with yellow to green skin and sometimes with a red blush. It is bitter when eaten raw but once cooked makes a pleasant preserve or jelly. Flowers and fruit are borne on old wood. There are several cultivars and some hybrids of each species. The plant is propagated by root cuttings, seeds sown outdoors in autumn as soon as they are ripe or stratified with cold treatments and sown in spring, or by semihardwood cuttings, by layering, and sometimes by grafting onto *C. speciosa* or *Cydonia oblonga* stocks. The plants are also easily divided. Cuttings taken in August, treated with 1,000 ppm IBA and struck in peat:perlite mix under mist will root easily (Dirr, 1998). The plants can be grown in almost any soil and in full sun and need only occasional corrective pruning. The fruit is harvested in early fall when its grassy-green undercolor has turned to yellow-green or yellow.

quince, Japanese *(Chaenomeles speciosa): See* **quince, flowering.**

quince, maules *(Chaenomeles japonica* var. *maulei): See* **quince, flowering.**

quonderberry: *See* **huckleberry, garden.**

rabbitberry *(Shepherdia argentea): See* **buffaloberry.**

rabbit thorn: *See* **wolfberry.**

raisin, wild: *See* **viburnum, blackhaw.**

raspberry: Rosaceae. BRAMBLE. There are hundreds of species in the genus *Rubus*. These are roughly divided into the general categories of "blackberries" (*see* **blackberry**) and "raspberries," with many hybrids of the two. In general, the blackberries are more or less thorny plants bearing fruit the core (receptacle) of which detaches from the plant and remains with the fruit after harvest. The raspberries form plants that are more or less free of vicious thorns and bear fruit the core of which separates from the fruit and remains with the plant after harvest. This makes their fruit "thimblelike." The plants of raspberries form woody shrubs with biennial canes. The first year canes are called "primocanes" and normally do not fruit. The second year canes are the normally fruiting "floricanes." Primocanes and floricanes may have different foliage, which is key to distinguishing the species.

Rubus deliciosus Torr. ROCKY MOUNTAIN FLOWERING RASP-BERRY, ROCKY MOUNTAIN RASPBERRY. This forms an unarmed upright bush about 2 meters (6.5 feet) in height. Leaves with three or five broad lobes; flowers rose-like, white, and mostly solitary; fruit is dark purple or wine-colored. The species is native to Colorado and is grown mostly as an ornamental.

Rubus idaeus L. EUROPEAN RASPBERRY, EUROPEAN RED RASP-BERRY, FRAMBOISE, RED RASPBERRY. This species forms an erect, more or less prickly plant about 1 to 2 meters (3-6.5 feet) in height. Leaflets three or five, gray or white pubescence beneath; flowers white and small; fruit conical, mostly dark red, sometimes yellow. Native to Eurasia and hardy to zone 4. All of the pomological red raspberries belong to this species and its varieties.

Rubus idaeus var. *canadensis* Richardson. The plant is native from Labrador to Colorado and north to Alaska.

Rubus idaeus var. *strigosus* (Michx.). AMERICAN RED RASPBERRY. This is hardier than the parental type and bears light red, soft fruit. It is found in the wild from Newfoundland to North Carolina, west to Wyoming and north to British Columbia, and in East Asia.

Rubus × *neglectus* (*Rubus idaeus* × *Rubus occidentalis*). PURPLE RASPBERRY. This plant resembles *Rubus occidentalis* in growth habit.

Rubus niveus Thunb. HILL RASPBERRY, MYSORE RASPBERRY. This plant forms canes to 2 meters (6.5 feet) in height. Leaflets five or seven; flowers small, purple; fruit red. The plant is native to India and Western China and hardy to zone 7.

Rubus occidentalis L. BLACKCAP, BLACK RASPBERRY, THIMBLE-BERRY. An erect plant up to 2 meters (6.5 feet) in height with prickly and glaucous canes. Leaflets mostly three; flowers small, white and in prickly clusters; fruit hemispherical, black or sometimes amber, firm. The plant is native from Quebec to eastern Colorado, south to Georgia and Arkansas and hardy to zone 4. This species is a parent for many black raspberry cultivars.

Rubus parviflorus Nutt. JAPANESE RASPBERRY, SALMONBERRY, THIMBLEBERRY, WESTERN THIMBLEBERRY.

Rubus phoenicolasius Maxim. WINEBERRY. This plant bears long canes with weak prickles and reddish brown glandular hairs, giving the canes a distinctly hairy appearance. Leaflets usually three; flowers small, white to pink, in tight clusters; fruit bright red, small and edible. The species is native to China and Japan and since its introduction into the United States has escaped cultivation. The plant is reportedly hardy to zone 6 though the authors have seen them growing rampantly in protected areas of zone 3.

Rubus rosifolius Sm. MAURITIUS RASPBERRY. A more or less trailing plant with canes up to 2 meters (6.5 feet) in length. Leaflets five or seven; flowers white, solitary or in clusters; fruit globose to oblong and red. This species is native to East Asia and hardy to zone 7. The fruit is edible and the roots are used in medicine.

Rubus spectabilis Pursh. SALMONBERRY. The long upright stems eventually become prostrate. They have peeling bark and few, small spines. Leaflets three; large flowers rose-colored; fruit salmon-colored and edible. This species is native to Idaho and California and north to Alaska and is hardy to zone 6.

Although the fruit of all raspberries is edible, that of many species is not palatable. The three main types of cultivated raspberries of

North America are derived from two species *(R. idaeus* and *R. occidentalis)* and a hybrid *(R. idaeus* × *R. occidentalis)* and account for about 75 percent of the total U.S. production. The United States produced 50,000 metric tons (55,115 tons) of all raspberries in 2004 (http://www.faostat.fao.org). Washington led U.S. production of red raspberries with 30,150 tons (27,351 metric tons) of fruit, followed by Oregon, with 3,350 tons (3,339 metric tons) (http://www.jan .mannlib.cornell.edu). Oregon also produced 1,100 tons (997 metric tons) of black raspberries. Worldwide, the Russian Federation led other countries in raspberry production with 170,000 metric tons (187,392 tons) followed by Serbia and Montenegro with 90,000 metric tons (99,208 tons) (http://www.faostat.fao.org). The United States followed as the third largest producer.

Humans have harvested wild raspberries for thousands of years, but the first recorded harvest of these fruit was reported to be on Mount Ida in Greece in AD 45, hence the species name *idaeus,* which means "pertaining to Ida." No raspberries grow there now, which leads botanists to believe the name originally referred to the Ides Mountains in Turkey. The Romans cultivated the raspberries for their fruit and for medicinal purposes and they were early introduced into the United States, where as many as four cultivars were offered for sale at the end of the eighteenth century.

Raspberries bear fruit on biennial canes, while the roots of the plants are perennial (for the most part) and annual. There are two types of bearing habits. In the June-bearing type a primocane is produced the first year, setting flower buds in late summer. Most fruit buds are located 0.5 to 1.5 meters (1.6-5 feet) off the ground, with the number of flowers contained in each bud partially a function of plant vigor. Larger canes produce more flowers than smaller canes (Bartels et al., 1988). The buds remain dormant over the winter and bloom the following summer, producing a crop on the floricane. Following fruiting, the cane dies and is replaced by new primocanes arising from the base of the plant. In the fall-bearing, or everbearing type, the primocane changes to a floricane when it has produced about a dozen buds. At the end of the first season the buds on the upper 20 centimeters (8 inches) or so of the cane bloom to form the fall crop. Following fruiting, the upper parts of the cane die. Buds lower down on the cane remain dormant, overwinter, and flower in the spring of the second year. Following this crop, the cane dies, to be replaced by primocanes arising from the plant's base. The fruit of red and yellow raspberries (the yellow usually is simply a color variation of the red) is borne along the single cane, whereas fruit of the black and purple raspberry is borne on laterals.

New cultivars of raspberries are being developed all the time and may be especially adapted to certain areas of the country. Cultivars of red raspberries recommended for the northeastern United States include the June-bearing 'Latham', 'Hilton', 'Taylor', and 'Newburg' and the fall-bearing 'Fallred' and 'Heritage'. 'Canby', 'Fairview', 'Meeker', and 'Willamette' do well in the northwestern United States. For the central United States, 'Dormanred'; for the central Atlantic coast of the United States, 'Reveille' and 'Pocahontas'. In Canada, 'Boyne' and 'Madawaska' perform well. The yellow everbearing cultivars 'Fallgold', 'Forever Amber', and 'Kiwigold' are hardy and widely adapted. Cultivars of black raspberries include 'Allen' and 'Jewel' for the eastern United States and 'Black Hawk' and 'Munger' for the western United States. The cultivars 'Clyde', 'Sodus', 'Royalty', and 'Brandywine' are outstanding purple raspberries.

Raspberries are both wind-pollinated and pollinated by honeybees. They are mostly self-fruitful but will produce better crops when provided with cross-pollination (Bartels et al., 1988). Red raspberries and most yellow raspberries are propagated by suckers from the crowns or roots. Black raspberries and most purple raspberries produce no underground shoots and so are best propagated by tip layering in late summer. Once rooted, the daughter plants may be removed later that autumn or the following spring (Bartels et al., 1988).

The best raspberry soils are deep and well drained but retentive of moisture, though black raspberries have some tolerance for lighter, sandy soils. Roots of all brambles begin to die in soils waterlogged for more than 24 hours, so be sure soil drainage is adequate and that the subsoil is not tight. Soils should be slightly acid and have about 4 percent organic matter. In general, raspberries do not tolerate high temperatures during the growing season and do poorly when temperatures rise above 27°C (80°F). They have optimal growing temperatures of 15.5 to 24°C (60-75°F) and require about 800 hours of chilling. Red raspberries are hardier than either black or purple raspberries and can tolerate winter temperatures as low as –29°C

(–20°F) when fully dormant. Strong winds will result in cane damage, so choose a protected site or construct a windbreak to the windward side of the planting. Destroy all wild brambles within a quarter mile of the planting to reduce the spread of disease and do not plant brambles on sites where eggplants, peppers, tomatoes, potatoes, strawberries, or other brambles have grown within the previous five years (Bartels et al., 1988).

Raspberries are usually planted in spring using one-year-old plants. Red raspberry plants are set about 1 meter (3 feet) apart in rows 3 meters (10 feet) apart, with spacing in part depending upon the size of equipment used in maintenance. Spacing for home plantings can be much closer. Black and purple raspberries, being larger plants, are spaced about 1.5 meters (5 feet) apart in rows about 3 meters (10 feet) apart. Red and yellow raspberries are pruned in the early spring by removing canes that had fruited the year before and damaged canes, and by thinning out floricanes to stand about 18 per linear meter of row (6 per linear foot). Floricanes longer than about 1.6 meters (5 feet) in height may also be headed back at this time by up to a quarter of their total length without much damage to the crop. The dead tops of fall-bearing canes that fruited the previous autumn should also be removed at this time. The primocanes of black and purple raspberries are headed back in midsummer to about 0.6 meter (2 feet) to stimulate formation of strong laterals. Early the following spring the laterals are headed back to about 20 centimeters (8 inches) to reduce overbearing. The fruited, dead canes are removed the following spring, to be replaced by emerging primocanes (Crandall, 1995; Crandall and Daubeny, 1990). The canes of red raspberries may be held in place by erection of a trellis system consisting of stout posts extending out of the ground about 1.8 meters (6 feet) and fitted with a cross arm about 0.5 meter (1.5 feet) in length. Wire run between the ends of the cross arms and along each side of the row will keep canes emerging beneath them in check. Do not fertilize plants within two months of planting. Then, apply nitrogen in the form of a complete fertilizer at the rate of about 47 grams (1.5 ounces) per plant. Fertilize early each spring with about 93 grams (3 ounces) of ammonium nitrate per 1 meter (3 feet) of row and, in the northern United States, never apply fertilizer after July 1 to reduce the incidence of winter damage (Pritts, 1996). Practice good weed control and pruning practices and a healthy planting of red raspberries can stay productive for 20 years. Black and purple raspberries will continue to produce well for about eight years.

Raspberry fruit is ripe about 80 to 90 days after bloom. The fruit is borne in loose clusters and is aggregate, with each seed surrounded by fleshy pulp. The receptacle remains firmly attached to the fruit until the latter is ripe. The fruit is harvested when it has attained its full color and begun to soften and when it separates easily from the receptacle. Overmature fruit turns dark and will drop or rot quickly. Harvest fruit of yellow raspberries when it has begun to turn an orange-salmon color. Placing the fruit in containers larger than a half pint will result in fruit on the bottom being crushed. Undamaged mature fruit stores well for a few days at temperatures between 0 and 4.5°C (32-40°F) and under reasonably high relative humidity. Raspberry fruit has a pH of 3.0 to 3.5 and a balanced ratio of sugars and acids. Fruit grown under warm, dry conditions is sweeter, less acid, and more highly colored than that grown in cooler, more moist conditions. Hot weather reduces fruit aroma and wet weather reduces fruit sugars. Raspberries contain good amounts of the anticarcinogen ellagic acid. They also contain small amounts of vitamins, though only vitamin C levels are significant.

raspberry, Japanese *(Rubus parviflorus): See* **raspberry.**

Ribes: Saxifragaceae. Both the currants and the gooseberries belong to the genus *Ribes,* which contains about 150 species. All produce edible fruit, though the common cultivated forms are derived from only a few species. All are shrubs native to temperate regions of the northern hemisphere. They may be armed or unarmed. Leaves alternate, often clustered and deciduous; flowers small, inconspicuous in most species, single or in racemes containing a dozen or more flowers; fruit a berry with prickles, glandular hairs, or smooth. The dried petals and parts of the pedicel remain attached to the fruit. There are two fairly distinct groups within the genus which may be separated into the "currants" and the "gooseberries." However, the two groups are so similar in their culture that they can be discussed as one. The species grown for their edible fruit are included in the following:

CURRANT GROUP. Plants in this group are usually unarmed.

Ribes alpinum. ALPINE CURRANT. Grown mostly as an ornamental, although the fruit is edible.

Ribes aureum. GOLDEN CURRANT. The fruit of this plant is also sometimes eaten and makes a clear golden jelly, though the plant is mostly valued as an ornamental.

Ribes nigrum L. BLACK CURRANT, CASSIS, EUROPEAN BLACK CURRANT. This is a vigorous bush up to 2 meters (6.5 feet) high bearing smooth black fruit. It is hardy to zone 5 and its cultivars are widely grown for their fruit. World production of black currant was reportedly 582,976 metric tons (520,621 tons) in 1989. Of this, Poland produced 169,354 metric tons (166,680 tons); Germany, 132,086 metric tons (130,000 tons); the Soviet Union, 106,684 metric tons (105,000 tons); and Czechoslovakia, 38,392 metric tons (37,786 tons). Production in the United States was 46 metric tons (45 tons) and mostly centered in California and the Pacific Northwest. Other world production areas include New Zealand and Europe. The fruit is used mainly in jellies and other desserts.

Ribes nigrum × *R. divaricatum* × *R. uva-crispa* L. JOSTABERRY, PRUTERBERRY, YOSTABERRY. This is a new plant developed in Europe in the 1940s. Because of it brief existence its proper culture is still being researched. However, it does appear to be highly resistant to the white pine blister rust. The plants are not thorny and the ripe fruit is deep black, tart, and borne singly like gooseberries, though ripening is uneven, making harvest difficult. There is some U.S. production, mainly in eastern Washington and Idaho, and in Canada.

Ribes odoratum H. Eendl. BUFFALO CURRANT, CLOVE CURRANT, FLOWERING CURRANT, MISSOURI CURRANT. The shrub is tall, growing to 2 meters (6.5 feet), and unarmed. It bears fragrant yellow, clove-scented flowers and black, glabrous fruit. The plant is native from South Dakota and Minnesota south to Texas and Arkansas and hardy to zone 5. It is usually grown as an ornamental, although its fruit is edible.

Ribes sativum (Rchb.) Syme. COMMON CURRANT, GARDEN CURRANT, RED CURRANT, WHITE CURRANT. The plant grows up to nearly 2 meters (6.5 feet) and is unarmed. The fruit is red or white and very juicy. It is native to Western Europe and hardy to zone 5. This species is mainly cultivated in Washington, with 56.5 hectares (40 acres) in

1995, and in California. New Zealand also produces the fruit commercially.

World currant production totaled 873,584 metric tons (962,961 tons) in 2004 (http://www.faostat.fao.org). The Russian Federation led production with 396,000 metric tons (436,515 tons) followed by Poland with 192,000 metric tons (211,643 tons) and Germany with 148,000 metric tons (163,142 tons). U.S. currant production ceased to be reported in the 1980s.

The red currant was cultivated in Germany as early as the fifteenth century, primarily for medicinal use. By the sixteenth century it was being grown as a garden fruit in England and by 1778 several cultivars had been introduced. The black currant was also used for medicine and was readily available in Russian markets by the eighteenth century. The first records of the currant in the United States date from 1770, though it was probably introduced by the earliest settlers. The leaves, roots, and fruit of the black currant reportedly have medicinal value. Black currant juice, tea, and extracts have been used to treat sore throats (quinsy) and are very high in vitamin C. The leaves and buds are reported to have antiinflammatory value. Oils extracted from the leaf and flower buds have been used in cosmetic creams, lotions, and perfumes. The black currant seeds are considered to be a potential source of omega-3 and omega-6 fatty acids currently used to treat asthma, premenstrual syndrome, and arthritis (Tulloch and St. Pierre, 1996).

GOOSEBERRY GROUP. Plants in this group are usually armed with stout spines and prickles.

Ribes hirtellum Michx. HAIRY GOOSEBERRY. This shrub grows to 1 meter (3 feet) and has bristly branches with green or purple flowers. The fruit is purple or black and usually smooth. It is native from Newfoundland to West Virginia and South Dakota and hardy to zone 4. This is an important gooseberry species.

Ribes uva-crispa L. ENGLISH GOOSEBERRY, EUROPEAN GOOSEBERRY. The plant grows to 1.5 meters (5 feet) in height and bears strong, stout spines up to 1 centimeter (0.4 inch) long. The fruit is red, yellow, or green and pubescent. The plant is native to Europe but has escaped from cultivation and naturalized in some areas. The species is hardy to zone 5 and has given rise to many cultivars of gooseberries.

Gooseberry cultivation began in Europe around 1700 and by 1740 there were over 100 cultivars reportedly grown in England. Estimates suggest there are now nearly 5,000 gooseberry cultivars grown worldwide at the beginning of the twenty-first century.

In 2004, Germany led the world in gooseberry production with 90,100 metric tons (99,318 tons) followed by the Russian Federation with 60,810 metric tons (67,031 tons) and Poland with 20,000 metric tons (22,046 tons).

In the United States, Oregon produces most of the gooseberries, with smaller acreages in New York, Minnesota, North Dakota, and Washington.

The currants and gooseberries are used almost solely for processing into jams, juice, jelly, and compote. Black currants especially are prized for their distinctive flavor and in France are made into brandy and a liqueur, creme de cassis, made specifically from the cultivar 'Noir de Bourgogne'. Candies, flavored honey, and flavored dairy products such as yogurt are also made from black currants. Ripe gooseberries are eaten fresh or made into pies and jams as is the green fruit (Harmat et al., 1990).

The red and white currants and the gooseberries produce fruit at the base of one-year-old wood and on spurs on two- and three-year-old wood. Older wood is relatively less productive. Black currants produce most of their fruit in clusters of 8 to 30 on one-year-old wood. Flower buds differentiate in mid to late summer of the season previous to bloom.

Most cultivars of red and white currants and the gooseberries are self-fruitful; those of black currant and jostaberry are entirely or partially self-sterile, making planting of two or more cultivars of these plants necessary for good fruit set. Self-sterility varies by degree among cultivars. Bees usually pollinate the flowers in this genus, but the plants bloom very early in the spring, often during weather inclement for bee activity.

Cultivars of red currants include 'Red Lake', 'Cherry', and 'Perfection'. The new cultivar 'Viking' is rust resistant. Of the white currants 'White Imperial' is considered among the best, and 'White

Grape' is a very hardy cultivar. 'Golden Prolific' is an amber selection from the native *R. odoratum*, while 'Crandall' is a black-fruited cultivar. Of the black currants, the cultivar 'Baldwin' accounts for about 80 percent of world production, though 'Ben Lomand' and 'Ben More' are even more productive than 'Baldwin'. Three rust-resistant cultivars developed in Canada are 'Consort', 'Crusader', and 'Coronet'. The berries of 'Consort' are soft and must be hand-harvested. The American gooseberry cultivars 'Poorman' and 'Welcome' (both red-fruited, the latter so named because it is relatively unarmed) and 'Pixwell' (yellow-fruited) are excellent. While the European cultivars are larger fruited than the American, they are not as hardy and are more subject to disease. The newer cultivars are therefore likely to be hybrids of the American and European species. 'Josta', 'Jostaki', and 'Jostine' are cultivars of jostaberry.

Most currants and some gooseberries can be propagated by hardwood cuttings. Take cuttings 15 to 20 centimeters (6-8 inches) long in late autumn and either plant them immediately outdoors or hold them in refrigerated storage for callusing and spring planting. Plant them deeply enough so that only the top node extends above the soil. Placing them in a mist chamber after treatment with 0.8 percent IBA in talc will result in rooting in about two weeks. Semihardwood cuttings taken in mid-June and treated with 0.4 percent IBA in talc, are placed in a mist chamber where rooting occurs in a week to ten days (Tulloch and St. Pierre, 1996). Gooseberries are generally propagated by mound layering.

European gooseberries root poorly from cuttings and are generally propagated by simple layering or mound layering. Any well drained soil is satisfactory for growth of these plants though they by far do best on the heavier cool clay loams and silt loams. Soil pH should be slightly acid to neutral. The plants have a shallow root system located mainly in the upper 20 to 40 centimeters (8-16 inches) of soil and encompassing a diameter rarely more than 80 centimeters (31 inches) (Harmat, 1990). These plants do best on a cool site and will not tolerate hot areas and southern exposures. Defoliation and leaf damage will occur with temperatures in excess of 30°C (86°F). Such heat damage limits their cultivation in the southern states. They are one of the few fruit plants that perform well on a northern, northeast, or northwest exposure and require 120 to 140 days to ripen properly.

Since they are hosts for the white pine blister rust they must not be planted near any five-needled pines. Many states regulate the planting of *Ribes* and it would be wise to check with your local extension service or department of agriculture to determine the suitability and legality of planting these fruit in your area (Barney and Hummer, 2005).

Set dormant one- or two-year-old plants in very early spring or in autumn. Since the plants begin growth very early in the spring, autumn setting may be preferable. Set the plants about 2 meters (6.5 feet) apart in both directions (Harmat et al., 1990). Keep weeds down by frequent cultivation or by application of an organic mulch. This is particularly beneficial since these plants do best on a cool soil and will shed their leaves in hot areas in midsummer. Fertilize the plants with a complete fertilizer beginning in the second spring. The amount of fertilizer applied should be equivalent to 115 grams (3.7 ounces) of 10-10-10 per plant. Increase the fertilizer in the third spring to about 230 grams (7.4 ounces) of 10-10-10 or equivalent. In subsequent years the planting may need up to 300 grams (9.6 ounces) per plant. If shoot growth needs invigorating, apply the fertilizer in the early spring; otherwise, mid-fall applications are preferable. It is particularly important to keep black currants growing vigorously in order to maintain good production (Harmat et al., 1990; Barney, 1996).

Remove all canes older than four years old on currants and gooseberries and allow only about 12 canes to remain per bush. Remove outlying canes so that the bush is no more than 0.6 meter (2 feet) in diameter at the base. Head back particularly long gooseberry shoots to prevent them from layering.

Currants require about 60 days from bloom to harvest, while gooseberries ripen from 60 to 80 days from bloom. The fruit of currants ripens over a two-week period but holds well on the bush for a week or more after ripening. Black currants often display premature fruit drop, called "run-off," occurring about a month after bloom. This is usually attributed to some combination of inadequate pollination, disease, and/or unfavorable environmental conditions (Tulloch and St. Pierre, 1996). Currants for jelly are picked somewhat underrripe when their pectin content is still high. For other uses they are harvested fully ripe. Harvest red and white currants by picking the entire cluster (called a "strig") when the fruit has fully colored and begun to soften. The fruit is then shelled from the strig at the time of use or processing. Gooseberries, black currants, and jostaberries are harvested individually when they have developed the appropriate color for the cultivar and have just begun to soften. Gooseberries are usually harvested when they are green to red, while black currants and jostaberries are harvested when fully black. As their name implies, red currants are harvested when they are a clear red and white currants a clear, translucent white. Be sure to wear heavy leather gloves when harvesting gooseberries. Good yields of red and white currants and gooseberries can amount to from 5.5 to 11 liters (5-10 quarts) per plant, while a good yield of black currants may be half that amount. Fruit size is usually from 1.5 to 2.5 centimeters (0.6-1 inch) diameter, depending upon species and cultivar.

The fruit of *Ribes* species is generally tart and is used in jams, jellies, compote, conserves, preserves, wine, and for medicine. The leaves and buds of the black currant are used for herbal medicines. White currants are used in baby foods in Europe but are somewhat less flavorful than the red. Gooseberries can be eaten out of hand and in flavor and texture resemble mild grapes. Jostaberries can be processed like currants or eaten fresh like gooseberries (Barney, 1996). The fruit of all *Ribes* spp. is rich in vitamins A, B, and C, pectins, minerals, citric acid, and fructose (Harmat et al., 1990). In fact, black currants contain about 90 to 355 milligrams of vitamin C per 100 grams of fruit, up to four times more than that in the orange. Red currants contain from 16 to 65 and gooseberries, 14 to 40 milligrams of vitamin C per 100 grams of fruit (Tulloch and St. Pierre, 1996).

rose, brier *(Rubus coronarius): See* **blackberry.**

rose, guelder *(Viburnum opulus): See* **cranberry, highbush.**

rosehips: Rosaceae. ROSE BRIER. There are more than 100 species of plants in the genus *Rosa,* many of which are grown as ornamental shrubs for their flowers. None are grown specifically for their fruit, though the fruit of all can be harvested and utilized. The fruit of all roses has been utilized from time to time throughout history. The fruit of *Rosa canina* was widely gathered in ancient Europe for use in tarts. It was also mixed with wine and sugar to make a jelly. In China, the fruit of *R. centifolia* was used to scent tea. The North American Indians used the fruit of several species and the fruit of *R. villosa* was used as a dessert and conserve in Europe and Asia (Hedrick, 1919). The species are mostly native to temperate parts of the northern hemisphere and have been heavily hybridized by plant breeders. Leaves alternate, mostly odd-pinnate, deciduous or persistent; flowers single or in corymbs or panicles; fruit is a fleshy hip containing the hairy achenes, or true fruit. Since none are grown for their fruit and the horticultural literature is replete with directions on culture of roses, this entry will not deal extensively with the many species and their culture. One exception is the Japanese rugose rose which has particularly large hips.

Rosa rugosa Thunb. JAPANESE ROSE, RUGOSA ROSE, RUGOSE ROSE, SEA TOMATO, TURKESTAN ROSE. A shrub up to 2 meters (6.5 feet) in height with densely prickly canes. Leaflets five to nine; flowers rose to white and up to 7.5 centimeters (3 inches) in diameter; fruit a hip up to 2.5 centimeters (1 inch) diameter. The plant is native to China and Japan and hardy to zone 2. The Ainos of Japan made wide use of the fruit of this species (Hedrick, 1919).

Roses normally bear their flowers on one-year-old wood and are usually pollinated by bees. Inclement weather, high winds, or lack of pollinating insects may hinder fruit set.

Roses are easily propagated by taking 15 centimeter (6 inch) softwood cuttings from stems that have just finished flowering and placing them into the ground under a glass jar in a shady location. They will have rooted by the following spring and may be transplanted at that time. Roses may also be budded on a stock such as *R. multiflora* or *R. canina*.

All species do well in well drained, slightly acid soils and will not tolerate poor soil drainage. Roses do best in full sun and where they are protected from strong winds and full bright sunlight. They may be planted in early spring or early fall and spaced about as far apart as the bush is expected to grow in height. In early spring prune out all winter damaged wood, dead wood, and small twiggy growth.

Harvest the hips when they have fully ripened and turned a bright red color, though the color may vary from brick red to scarlet to orange, depending upon the species. Rosehips are high in vitamin C, and were used in the manufacture of that vitamin from time to time when usual sources were scarce. For example, during the Nazi blockade of Great Britain during World War II, school children gathered rosehips from the countryside for use in manufacturing vitamin C.

roundwood *(Sorbus americana): See* **ash, mountain.**

sallow thorn *(Hippophae rhamnoides): See* **buckthorn, sea.**

salmonberry *(Rubus parviflorus): See* **raspberry.**

sandberry *(Arctostaphylos uva-ursi): See* **bearberry.**

sarviceberry: *See* **serviceberry.**

sarvis: *See* **serviceberry.**

sarvis tree: *See* **serviceberry.**

Saskatoon *(Amelanchier alnifolia): See* **serviceberry.**

scuppernong *(Vitis rotundifolia): See* **grape.**

serviceberry: Rosaceae. ALLEGHENEY SERVICEBERRY, CHINESE SERVICEBERRY, CURRANT TREE, INDIAN PEAR, JUNEBERRY, MOUNTAIN JUNEBERRY, SARVIS, SARVICEBERRY, SARVIS TREE, SASKATOON, SHAD, SHADBERRY, SHADBLOW, SHADBUSH, SNOWY MESPILUS, SUGARPLUM, SWAMP SUGAR PEAR. All of these names, especially "serviceberry," are applied to several members of the genus *Amelanchier,* comprising about 25 species of shrubs and small trees native to the northern temperate zone, though in some parts of the world the term "serviceberry" is applied to members of the *Sorbus* genus. Leaves alternate and toothed; flowers white and borne mostly in terminal racemes or open clusters in early spring; fruit is a small dark purple or black pome about 8 millimeters (0.3 inch) in diameter and edible in most species (Hortus, 1976; Jones, 1946; Miller and Stushnoff, 1971).

Amelanchier is extremely winter hardy and tolerates dry, alkaline conditions of the Great Plains up into the prairie provinces of Canada and as far north as southern Yukon and Northwest Territories (Gough, 2002; Laughlin et al. 1988). The fruit of all species has a heavy bloom and fairly insipid flavor, making it better for preserves than for eating out of hand. Plants are also useful in the landscape for their early showy flowers, which usually open well before foliation.

The term "sarvis tree" is one of the oldest common names for this plant and apparently represents a confusion by the early English settlers with *Sorbus domestica,* the common name for which was formerly "sarvis" or "service tree."

Amelanchier alnifolia Nutt. KOREAN JUNEBERRY, MOUNTAIN JUNEBERRY, ROCKY MOUNTAIN BLUEBERRY, SASKATOON, WESTERN SERVICEBERRY, WESTERN SHADBUSH, WILD PEAR. This species is native to the southern Yukon and Northwest Territories south to the northern prairie states, Oregon, and Idaho. It is the species from which most domesticated clones have been developed. It grows under a wide range of soil and climatic conditions but does poorly on heavy, poorly drained clay soils low in organic matter. The plant is a shrub or shrubby tree growing to heights of about 6 meters (19.6 feet) and bears masses of white flowers. The fruit is borne in clusters of six to twelve and ranges in color from blue-black to cream to red. Individual fruits from wild plants may reach sizes of up to 1 centimeter (0.4 inch) diameter, while those of cultivated plants may reach 1.6 centimeters (0.6 inch) diameter and are quite sweet. There is very little commercial production in the United States, though Canada reported about 345 hectares (850 acres) in production in 1996. The name is derived from the Native American name "mis-sask-quah-too-min." It is a prized fruit in the northern plains and was extensively used by the Native Americans in pemmican. Today the juicy, low-acid fruit is eaten fresh, made into wine, or baked into pies and other desserts. The fruit, usually referred to as a "berry" is botanically a pome, like apple and pear, to which it is related. The taste of the processed fruit is enhanced by addition of lemon juice (Hedrick, 1919). Several species are used for game range restoration and wildlife plantings, for windbreak plantings, and for low maintenance or native plant landscaping (Laughlin et al., 1988).

All species bear fruit on one-year-old and older wood and plants usually produce their first crop two to four years after planting (Gough, 2002). The plants bloom very early in spring, making them particularly susceptible to spring frosts. 'Altaglow' (self-sterile but producing very sweet, white fruit), 'Shannon', and 'Indian' are superior cultivars with large fruit. 'Pembina' and 'Smoky' are also productive but produce slightly smaller, very sweet fruit. 'Forestburg' produces the largest fruit, which matures slightly later than that of other cultivars, but fruit flavor is inferior to that of other cultivars.

Seed is slow to germinate. Sow them in autumn. Transplanting is the usual but difficult method. For the best success, transplant plants 46 to 61 centimeters (18-24 inches) tall in early spring, cutting them back to 15 centimeters (6 inches) above the ground at planting. Take care to preserve as many fine roots as possible. Set the plants 1.8 to 2.4 meters (6-8 feet) apart each way (home plantings) or on 2 meter (6.5 foot) centers with rows spaced 4 meters (13 feet) apart (commercial plantings), and about 5 to 8 centimeters (2-3 inches) deeper than they grew originally (Patterson, 1957; Laughlin et al., 1988). Crown division of plants at least five years old will provide between 10 and 25 transplant divisions from each plant (Laughlin et al., 1988). Transplanting suckers is also productive. Do this in early spring, cutting the tops off about 5 centimeters (2 inches) above the roots. Keep the plants lightly shaded until they reestablish a strong root system. Root cuttings about pencil-thick and 10 to 15 centimeters (4-6 inches) in length taken from dormant plants in the early spring and planted immediately will develop into new plants quickly. Be sure the stem end of the root is pointing upwards slightly and is buried no more than about a quarter inch deep. Pretreatment of root cuttings by placing them in poly bags filled with damp peat moss and storing them in the dark for three weeks at 21°C (70°F) promotes sprouting and faster rooting. Shoots will become obvious in about three weeks at which time the plants are placed in a seeding flat and covered with a light, nonsoil medium in the greenhouse. Allow the shoots to grow for two more weeks, then remove the plants and root them in moist medium (Laughlin et al., 1988). Softwood cuttings about 10 centimeters (4 inches) in length will root easily if taken in spring during active growth. Dip them in about 0.6 percent IBA talc and strike them in a mist bed with bottom heat of about 21°C (70°F). They will be rooted well enough for transplanting to the field the following spring. Some cultivars are also compatible with and are easily propagated by grafting or budding onto rootstocks of *Sorbus aucuparia* L. and *Coto-*

neaster acutifolia (Hilton, 1982). A principal difficulty in advancing the planting of this crop is our inability to propagate the large-fruited clones efficiently and in sufficient numbers to meet the demand.

Plants will form interspecific hybrids between *A. alnifolia* and male parents of crabapple, pear, and mountain ash (*Sorbus* ×*alnifolia*). These progeny are under trial in Alberta, Canada (Stang, 1990).

Amelanchier is not fussy about its soil requirement and does well in all soil types at pH from 6.0 to 7.8, except on poorly drained or poorly aerated tight clay soils lacking sufficient humus (Harris, 1972). The plants bloom relatively early in spring and so may be subject to late spring frosts. Select sloping sites with good air drainage. Plant dormant, well branched two-year-old plants in early spring on spacings of about 3 meters (10 feet) in both directions to permit cross cultivation, or 2 meters (6.5 feet) apart in rows 4 meters (13 feet) apart for hedgerow plantings. Prune off one-third of the top growth at planting to reduce water loss. Serviceberries in home plantings and established plants in commercial plantings require little irrigation, although a couple of irrigations each season will prove beneficial on drier sites and will improve berry size (Laughlin et al., 1988). Fertilizer requirements have not been established and may vary widely by location and soil types. Fertile soil often will not require supplementary fertilizer. On soils with lower fertility band nitrogen at the rate of 28 grams (1 ounce) of N per plant per year of plant age up to a maximum of 227 grams (0.5 pound). Avoid inducement of excessively vigorous growth where fireblight is a problem.

Plants of this species begin to bear when they are two to four years old. Fruit is produced on previous year's growth and on older wood, with the young, vigorous branches yielding the best crops. Prune the plants in early spring before growth begins, heading back the larger canes so that bushes are no taller than about 2 meters (6.5 feet). Remove all weak, diseased, and damaged growth and low branches and thin the center of the plant to keep it open. Remove all suckers ranging outside a 1 meter (3 feet) diameter crown. Follow renewal pruning on old bushes to improve production.

The fruit of this plant ripens evenly about 38 days after petal fall (Laughlin et al., 1988) and the whole crop can often be harvested one fruit at a time in a manner similar to that used for highbush blueberry.

Yields of over 2.4 metric tons per hectare (6 tons/acre) are not uncommon (Harris, 1972), with individual mature plants producing up to about 4.5 kilograms (10 pounds). Birds can be a real problem, feeding heavily upon the ripening fruit. Additionally, spring frosts may ruin the blossoms in some years, reducing production drastically. The fruit is used today in making jams, pie fillings, and fruit leathers. They contain more protein, fat, fiber, calcium, magnesium, manganese, barium, and aluminum than blueberries and strawberries and are considered a good source of manganese, magnesium, and iron (Laughlin et al., 1988).

Amelanchier canadensis (L.) Medic. GRAPE-PEAR, SWEET PEAR. This species is fastigiate, growing up to 8 meters (26 feet) in height, is native in swamps from Quebec to Georgia and hardy to zone 4. The fruit is pea-sized, edible, and used in a manner similar to those of *A. alnifolia*.

Amelanchier florida Lindl. FLORIDA JUNEBERRY. A shrub or small tree up to about 10 meters (30 feet). The fruit is nearly black with little bloom. The plant is hardy to zone 2 and was introduced in 1826. The species is similar to *A. alnifolia*.

Amelanchier oblongifolia Roem. SWAMP SUGAR PEAR. This is an upright shrub reaching a height of about 6 meters (18 feet). Cultivated since 1641 for its black, bloomy, sweet fruit. The plant is hardy to zone 4.

Amelanchier ovalis Med. EUROPEAN JUNEBERRY. Small spreading shrub up to about 2.5 meters (8 feet) and hardy to zone 4. The plant was introduced in 1596 and is grown for its blueish-black, bloomy fruit.

Amelanchier sanguinea (Pursh) DC. This is a straggly shrub 1 to 3 meters (3.2-10 feet) high found from Quebec to Ontario and south to North Carolina and Iowa. It is hardy to zone 5 and produces fruit about 0.5 centimeter (0.2 inch) diameter.

Amelanchier spicata (Lam.) C. Koch. A stoloniferous species growing to 2 meters (6.5 feet) in height, it is found from Quebec to Ontario and south to Pennsylvania, Ohio, and South Dakota. Hardy to zone 4. Some authors believe it to be a hybrid of *A. canadensis* × *A. ovalis*.

Amelanchier stolonifera Wieg. QUEBEC BERRY, RUNNING SERVICE-BERRY. This species also is hardy to zone 4 and found native from Newfoundland to Ontario and south to Virginia, Michigan, and Minnesota. The fruit is edible. 'Success' is one useful cultivar.

serviceberry *(Sorbus americana): See* **ash, mountain.**

service tree, wild *(Sorbus torminalis): See* **chequers.**

shad: *See* **serviceberry.**

shadberry: *See* **serviceberry.**

shadblow: *See* **serviceberry.**

shadbush: *See* **serviceberry.**

sheepberry: Caprifoliaceae. BLACK HAW, COWBERRY, NANNY-BERRY, NANNY PLUM, SWEETBERRY, SWEET VIBURNUM, TEA PLANT, WILD RAISIN. This deciduous shrub, *Viburnum lentago* L., grows up to 10 meters (30 feet) in height. The small, sweet, blue-black fruit is an edible single-seeded drupe about 12 millimeters (0.5 inch) long and borne in loose clusters which ripen about 90 to 110 days after bloom. There is very little U.S. production and world production is unknown. Culture is as for other viburnums. *See* **cranberry, highbush.**

silverberry: Eleagnaceae. The common term "silverberry" refers to both *Elaeagnus angustifolia* and *E. argentea,* the latter now classified as *E. commutata* Bernh. This latter species is a deciduous shrub up to 4 meters high and native to eastern Canada and south to Minnesota and Utah, being hardy to zone 2. The plant produces a mealy, dry, edible drupe about the size of a small cherry (Hortus, 1976; Hedrick, 1919). For culture, *see* **olive, autumn** or **olive, Russian.**

skunkbush *(Rhus trilobata* var. *malacophylla): See* **squawbush.**

sloe *(Prunus americana): See **Prunus.***

solanberry *(Solanum melanocerasum): See* **huckleberry, garden.**

sourtop *(Vaccinium myrtilloides): See* **blueberry** or **whortleberry.**

squawbush: SKUNKBUSH. This foul-smelling bush, *Rhus trilobata* var. *malacophylla* (Greene) Munz, grows up to 2 meters (6.5 feet) in height. Leaves are compound; flowers greenish and formed in clustered spikes. The red, hirsute fruit was once used by the Native Americans. This plant is a close relative of the sumacs and of poison ivy. Therefore, be sure you identify the plant positively before using the fruit.

squawbush *(Viburnum trilobum): See* **cranberry, highbush.**

stagberry *(Viburnum prunifolium): See* **cranberry, highbush.**

strawberry: Rosaceae. About 12 species of plants make up the genus *Fragaria,* from the Latin word *fraga,* meaning the "scent of a berry." All are low perennial stoloniferous herbs native to the northern temperate regions. They have a short, thickened "stem," called a crown, which has a growing point at its upper end and forms roots at the bottom. Under good growing conditions, a branch crown also develops from axillary crown buds, adding to the total yield. Leaves are compound; flowers white or pinkish, borne on cymose or racemose scapes; fruit is not a true berry but the usually red fleshy receptacle that bears the seedlike achenes, the true fruit. As a result of vigorous breeding efforts this plant can now be grown throughout North America and has the widest distribution of any temperate zone fruit.

The strawberry in general was widely mentioned by Roman writers, though the Greeks and Egyptians seem to have been unfamiliar with the plant. Nor was the fruit mentioned in English cookbooks up through around AD 1500. Its culture in European gardens was mentioned in the mid-1500s at which time it was also mentioned as growing wild in shady areas. Cultivated strawberries are often mentioned in the literature of the latter sixteenth century (Hedrick, 1919).

Fragaria ×*ananassa* Duchesne. CULTIVATED STRAWBERRY, GARDEN STRAWBERRY, PINEAPPLE STRAWBERRY. The fruit of this hybrid, *F. chiloensis* × *F. virginiana,* may grow to 2.5 centimeters (1 inch) in length. All contemporary cultivars of this hybrid are perfect-flowered. The United States led the world in fresh strawberry production in 2004, reporting a harvest of 1,004,110 metric tons (1,106,720 tons). Spain ranked a distant second with 286,000 metric tons (315,261 tons) followed by the Russian Federation with 215,000 metric tons (236,996 tons) (http://www.faostat.fao.org). In 1995, California reported 9,550 hectares (23,600 acres); Florida, 2,430 hectares (6,000 acres); Oregon, 2,300 hectares (5,700 acres); New York and North Carolina, 970 hectares (2,400 acres) each; Michigan, 730 hectares (1,800 acres); Pennsylvania, 570 hectares (1,400 acres); and Washington, 525 hectares (1,300 acres) of strawberries in production (Markle et al., 1998). This is also a highly popular garden plant and is planted throughout the country. Other world producing areas include Canada, with 5,750 hectares (14,214 acres), and New Zealand, Mexico, and Guatemala.

Early New World explorers brought back to Europe both *Fragaria virginiana* and *Fragaria chiloensis*. These were widely cultivated and chance crosses between the two species in England and Europe in the mid-eighteenth century resulted in this hybrid. In 1766, the Dutch horticulturist Antoine Nicholas Duchesne recognized that the hybrid combined the size and firmness of the Chilean strawberry with the flavor and productivity of the Virginia strawberry (Poling, 1996; Galletta and Bringhurst, 1990). This hybrid fruit is now the leading small fruit grown in temperate regions around the world, with large commercial industries in California, Florida, Mexico, Japan, Poland, Italy, Spain, and Russia. Most cultivars are self-fruitful (except 'Apollo'), though cross-pollination will result in bigger and better fruit. Strawberries are mostly wind pollinated though bees contribute to the process and can increase berry size by 20 percent and reduce the amount of malformed fruit by more than 25 percent. Blowflies and syrphids also contribute to the pollination process (Poling, 1996). There are three types of bearing habits in strawberry: the June bearing, the everbearing, and the day-neutral.

Growth in the June bearers is strongly affected by temperature and day length, with runner production occurring during long days and warm temperatures. Flower buds are formed under the short day, cool conditions of the early autumn. The buds overwinter and flower early the following spring, with berry ripening beginning about a month after bloom. Fruit quality of this bearing type is considered the best.

The everbearing type produces crops in spring and late summer, but the combined yield of both harvests generally does not equal the yield of the single harvest of a June-bearing cultivar. Flower buds that set late in the season overwinter and bloom early the following spring, producing the first crop of the year. Flower buds that form during the summer, under long day conditions bloom and produce fruit in late summer. The everbearers originated in the northern United States and Canada and so usually do best in those areas. Since hot weather stops flower bud development they generally do poorly in southern areas.

The day-neutrals can form flower buds during any length of day and in any part of the growing season provided the temperatures do not exceed 27°C (80°F). These have bearing habits similar to those of the everbearers, except that they produce continuously throughout the summer (provided the temperature is cool enough) and not simply in early and late summer. The fruit quality of some cultivars, especially 'Tristar' and 'Tribute' is considered equal to that of some June bearers and better than any of the everbearers (Poling, 1996). The strawberry is highly regional in its adaptation, with certain cultivars performing best within their given region. Also, like the peach among the tree fruit and the tomato among the vegetables, many new strawberry cultivars are frequently released. This makes any list of cultivars outdated almost as soon as it is printed. Always consult your local county extension service for the latest strawberry cultivar recommendations for your area. However, some cultivars are generally adapted to various regions and have proven themselves over many years of trials. These appear in the following:

June-bearing cultivars:

Northeastern United States: 'Allstar', 'Kent', 'Raritan', 'Redchief', 'Catskill', 'Earlidawn', 'Midway', and 'Surecrop'

Mid-Atlantic States: 'Allstar', 'Earliglow', 'Titan', 'Marlate', and 'Surecrop'

Gulf States, including Florida: 'Allstar', 'Earliglow', 'Oso Grande', 'Sweet Charlie', 'Earlibelle', 'Florida Ninety', and 'Headliner'

Central United States: 'Guardian', 'Honoeye', 'Redchief', 'Atlas', 'Delite', 'Midway', and 'Surecrop'

Pacific Northwest: 'Benton', 'Hood', 'Northwest', and 'Rainier'

California: 'Aliso', 'Heidi', and 'Tioga'

Across the northern United States, including the Plains States: 'Badgerbell', 'Catskill', 'Midway', 'Redcoat', 'Sparkle', and 'Surecrop'

Nova Scotia: 'Acadia' and 'Bounty'

Ontario: 'Veestar' and 'Vibrant'

British Columbia: 'Cheam', 'Northwest', and 'Totem'

Everbearing cultivars:
   Eastern United States: 'Gem', 'Ozark Beauty', 'Streamliner', and 'Superfection'
   Midwestern United States: 'Arapahoe', 'Gem', and 'Ogallala'
   Northwestern United States: 'Nisqually' and 'Quinalt'
Day-neutral cultivars:
   'Tristar', 'Tribute', 'Fern', and 'Selva', with the latter two being well adapted to California and Florida

New plants are formed on runners during the growing season. The abundance of runners is somewhat cultivar dependent, with some cultivars producing many and some few runners. Runners produce daughter plants, which are separated from the mother plants and used to establish a new planting. Well drained soil will reduce the chance of soil-borne diseases in strawberries and therefore most cultivars do better on somewhat sandy soils. Nevertheless, the soil should be fertile and have a good supply of moisture throughout the growing season.

This fruit has a shallow root system and so cannot tolerate drought for any length of time. Large commercial plantings may practice soil fumigation to control nematodes and other pests. This is not practical in home plantings where a cover crop may be rotated between plantings to reduce pests and to increase soil fertility. Because the strawberry is susceptible to *Verticillium* wilt, do not set a new planting in soil previously planted to eggplant, tomato, potato, or peppers, or other strawberries within the last five years. Locate the strawberry planting in full sun and outside of frost pockets, which are particularly troublesome to this early-flowering fruit.

In the north, strawberries are usually planted in early spring, while autumn planting is best in California and the south. Set the crowns level with the soil surface.

Strawberries quickly can grow out of bounds by natural simple layering. The swollen ends of the stolons root easily and produce runner plants. Each runner plant, called the "daughter" plant, produces runners, which in turn produce more daughter plants. Long days and warm temperatures promote the production of runners. Different cultivars also have an inherent capacity to form more, or fewer, runners (Poling, 1996). After setting the plants, remove all the flowers on the day-neutrals for the first few weeks of the season. On everbearers remove the flowers that will produce the first crop of the season, then let the plants produce normally for the second crop. Traditionally, all flowers on the June-bearing cultivars were removed the first year and the plants allowed to produce normally in subsequent years, but it is a more common practice today to remove flowers on these plants only for the first few weeks. The purpose of flower removal is to give the plant time to become established before having the stress of producing a crop.

The size of the bed and its design depends upon the use to which the planting will be put. Many home plantings and commercial plantings, except in California and Florida, utilize the matted row system in which plants are set about 0.6 meter (2 feet) apart in rows spaced about 1.5 meters (5 feet) apart. In California and Florida strawberries are often grown as an annual crop, with spring harvest following a fall planting. In the matted row system, spring-planted runner plants are allowed to grow and form daughter plants, which are then confined to a bed about 0.6 meter (2 feet) wide. Plants in this system produce the most fruit, but fruit size is somewhat smaller than in some other systems. Also, plant competition is great and weed control is difficult. In the hill system, often used for everbearers and day-neutrals, plants are spaced about 0.3 meter (1 foot) apart in single or multiple rows. All runners are removed and no daughter plants are allowed to develop. The mother plants become quite large, often developing multiple crowns. They are highly productive and fruit size is large. Also, weeding is relatively easy. The last training system is the spaced row system, wherein plants are set about 0.6 meter (2 feet) apart and a set number of runners (often four) are spaced an equidistance apart, held in place by pegs, small stones, or hairpins, and allowed to form daughter plants. All additional runners are removed. Plants in this system form a moderate amount of moderate-sized fruit. Weed control is easier than in the matted row system but a bit more difficult than in the hill system.

Winter protection is needed in many northern areas. The traditional method is to cover the beds with a few inches of straw after the soil has frozen in late autumn. Wheat or barley straw are preferred and one bale should be enough to cover approximately 9 square meters (100 square feet) of bed. Leaves and grass clippings are not suitable for this purpose since they mat and smother the plants. The winter mulch prevents the plants from heaving and also from desiccating where snowcover is insufficient or not reliable. The mulch is pulled off the plants when spring growth begins and placed into the walkways to act as summer mulch and to keep the fruit clean.

Nitrogen is the element most frequently needed in strawberry beds. About one and two-thirds teaspoon of ammonium sulfate per plant applied one month after planting will correct any obvious deficiencies. Apply fertilizer regularly at the end of the season at a rate equivalent to about 1.75 kilograms of ammonium sulfate per 30 linear meters (5 pounds/100 feet) of row. Fertilizer applied in the early spring often creates excessive vegetative growth that can cover the fruit and promote early rot. Also, the heavy canopy may interfere with adequate pollination.

Renovate an old bed or establish a new bed every several years to keep harvest productive. Begin renovation right after harvest by topdressing with a complete fertilizer such as 10-10-10 at the rate of 1.5 kilograms per 30 meters (4 pounds/100 feet) of row. With a rotary mower with the blade set to 10 centimeters (4 inches), mow the leaves from the plants. Rake and dispose of them. Reestablish walkways and narrow the beds with a rototiller, then thin the remaining plants to stand about 15 centimeters (6 inches) apart in the beds. Remove weeds and off-type plants. The season may not be long enough to allow for renovation in northern areas.

Harvest usually begins about a month after bloom. The first flowers to bloom, called "primary" flowers, produce the largest, though most "seedy," fruit. The secondary fruit ripens next and is next largest, then the tertiary and quaternary fruit. The fruit is borne on trusses, with three or more per truss. Fruit is 12 to 37 millimeters (0.5-1.5 inches) in diameter and generally concentric. Harvest only the red fruit. Since the fruit is highly perishable, harvest every couple of days, or every day in hot weather, and pick out and discard all rotted and damaged fruit each time to reduce the rate of decay in the crop. Separate the fruit from the plant with the hull attached and refrigerate immediately, unwashed. Wash and hull the fruit just before use. A single cup of berries contains about 45 calories. The fruit is an excellent source of vitamin C and is low in sodium.

Fragaria chiloensis (L.) Duchesne. BEACH STRAWBERRY, CHIL-EAN STRAWBERRY, FRUITILLAR, GARDEN STRAWBERRY, PINE STRAWBERRY. This is one of the parents of the garden strawberry. It forms mostly dioecious plants with a short, thickened rootstock. The fruit is rose-colored with white flesh. The species is native from coastal Alaska to California and South America. It was carried from Concepción, Chile, to Europe in 1712 (Hedrick, 1919).

Fragaria moschata Duchesne. HAUTBOIS STRAWBERRY. A large, white-fruited type was known as BOHEMIAN BERRY, CAPITON BERRY, or CAPRON. Flower types include perfect, pistillate, and staminate. The fruit has a musky flavor and is native to the forests and tall grass country throughout Europe east into Russia and Siberia.

Fragaria vesca L. ALPINE STRAWBERRY, FRAISES DES BOIS, PER-PETUAL STRAWBERRY, SOW-TEAT STRAWBERRY, WOODLAND STRAW-BERRY. The thick, woody rootstocks produce long, arching runners. The bisexual flowers are about 1 to 1.5 centimeters (0.4-0.6 inch) in diameter and produce mostly red, highly aromatic fruit up to 2 centi-meters (0.6 inch) in length. The achenes are raised. The species is na-tive to Eurasia and occurs now in the wild in North America, to which it was introduced from Europe. It is also found in the Andes of South America and on some Pacific islands. 'Alpine' and 'Perpetual' were derived from this species, which is the most widely distributed *Fragaria* species in the world. Prior to the introduction of *F. vir-giniana* into Europe in 1629, this was the species most often gathered from the wild, and is the species mentioned by Virgil, Ovid, and Pliny (Hedrick, 1919). Cato grew this fruit in his garden (Galletta and Bringhurst, 1990). Its culture in Europe became extensive during the fourteenth century and these fruit were the principle species of Euro-pean strawberry commerce until the nineteenth century.

Fragaria vesca L. ssp. *sempervirens*. ALPINE STRAWBERRY. This subspecies is thought to have originated in the mountains of Italy. The plant bears fruit almost continuously and is sometimes used as an edging plant along walks and gardens.

Fragaria virginiana Duchesne. VIRGINIA SCARLET STRAWBERRY, VIRGINIA STRAWBERRY. The plants are dioecious and form a thick rhizome. The flowers are up to 2.5 centimeters (1 inch) in diameter and bear fruit up to 2 centimeters (0.6 inch) diameter. These have red-dish pulp and the achenes are borne in deep pits. Plants are native from Newfoundland to Alberta and south to Georgia, Tennessee, and Oklahoma. This species is one of the parents of the garden strawberry and was taken to England early in the seventeenth century, being first mentioned there in 1629 (Hedrick, 1919).

sugarplum: *See* **serviceberry.**

summerberry *(Viburnum trilobum): See* **cranberry, highbush.**

sunberry *(Solanum melanocerasum): See* **huckleberry, garden.**

sweetberry: *See* **cranberry, highbush.**

sweet-hurts *(Vaccinium angustifolium): See* **blueberry.**

Swiss-cheese plant *(Monstera deliciosa): See* **monstera.**

T

tangleberry: *See* **huckleberry.**

tanglefoot *(Viburnum alnifolium): See* **cranberry, highbush.**

tanglelegs *(Viburnum alnifolium): See* **cranberry, highbush.**

tayberry: An 'Aurora' blackberry × raspberry hybrid selected by the Horticulture Institute in Scotland. *See* **blackberry.**

tea, kutai *(Arctostaphylos uva-ursi): See* **bearberry.**

tea, mountain *(Gaultheria procumbens): See* **wintergreen.**

teaberry *(Gaultheria procumbens): See* **wintergreen.**

thimbleberry *(Rubus occidentalis* or *R. parviflorus): See* **raspberry.**

tomatilla: *See* **wolfberry.**

tomatillo *(Physalis ixocarpa): See* **cherry, ground.**

tomato, cherry *(Physalis peruviana): See* **cherry, ground.**

tomato, gooseberry *(Physalis peruviana): See* **cherry, ground.**

tomato, husk *(Physalis ixocarpa): See* **cherry, ground.**

tomato, Mexican husk *(Physalis ixocarpa): See* **cherry, ground.**

tomato, sea *(Rosa rugosa): See* **rosehips.**

tomato, strawberry *(Physalis alkekengi): See* **cherry, ground.**

tree, beef-suet *(Shepherdia argentea): See* **buffaloberry.**

tree, cranberry *(Viburnum trilobum): See* **cranberry, highbush.**

tree, currant: *See* **serviceberry.**

tree, snowball *(Viburnum opulus): See* **cranberry, highbush.**

tree, whitten *(Viburnum opulus): See* **cranberry, highbush.**

triptoe *(Viburnum alnifolium): See* **cranberry, highbush.**

tuna: *See* **pear, prickly.**

umkokolo: *See* **kei-apple.**

veitchberry *(Rubus ursinus): See* **blackberry.**

viburnum, blackhaw: NANNYBERRY, SHEEPBERRY, SWEET HAW, WILD RAISIN. The fruit of *Viburnum pruni-folium* is a drupe up to about 1 centimeter long, pink to rose in color and turning black at maturity. It has been long used in preserves and jellies and the roots were used medicinally. The plant is hardy to zone

3 and was introduced into cultivation in 1727 (Dirr, 1998). Culture as for other viburnums. *See* **cranberry, highbush.**

viburnum, sweet: *See* **cranberry, highbush.**

vine, breadfruit *(Monstera deliciosa): See* **monstera.**

vine, silver *(Actinidia polygama): See* **kiwifruit.**

vine, tara *(Actinidia arguta): See* **kiwifruit.**

waiawi *(Psidium littorale): See* **guava.**

wallwort *(Sambucus ebulis): See* **elderberry.**

wayfaring tree, American *(Viburnum alnifolium): See* **cranberry, highbush.**

whimberry *(Vaccinium vitis-idea): See* **lingonberry.**

whinberry *(Vaccinium myrtillus): See* **whortleberry.**

windowplant *(Monstera deliciosa): See* **monstera.**

wineberry *(Rubus phoenicolasius): See* **raspberry.**

witch *(Viburnum alnifolium): See* **cranberry, highbush.**

witch-hobble *(Viburnum alnifolium): See* **cranberry, highbush.**

whortleberry: Ericaceae. BILBERRY, BLAEBERRY, SOURTOP, WHIN-BERRY. This shrub, *Vaccinium myrtilloides* L., grows up to 0.5 meter (1.5 feet) high. The flowers are solitary, short-pedicled and usually pinkish in color. The four to five celled fruit is sweet, black, covered with heavy bloom and about 8 millimeters (0.3 inch) in diameter. This species is somewhat more shade-tolerant than other *Vaccinium* species and is native to Europe, northern Asia, and northwest United States. The fruit is used for preserves and wine and also sometimes medicinally for treatment of cystitis. It is a favorite in Scotland. The plant was first cultivated in 1789 and is hardy to zone 4. For more information on its cultivation, *see* **blueberry**.

wintergreen: Ericaceae. There are about 100 species of evergreen shrubs in the genus *Gaultheria*. All are native to the Andes of South America but are also found in North America and from Asia to Australia. Leaves mostly alternate; flowers pink or white and borne singly or in racemes or panicles; fruit is a capsule, usually enclosed by a fleshy, brightly colored calyx (Hortus, 1976). Since they belong to the Ericaceae family these plants do best on moist, well drained acid soils high in organic matter. The site should have partial, light shade. They are widely grown as ornamentals on the west coast of the United States but none are cultivated primarily for their fruit. Plants are propagated by stratified seed, layers, division, and sometimes softwood cuttings (Wyman, 1986).

Gaultheria humifusa (R.C. Grah.) Rydb. ALPINE WINTERGREEN. This species forms a matted plant up to about 10 centimeters (4 inches) in height. The solitary white flowers give way to scarlet fruit that is used in preserves. The species is native from British Columbia to Colorado and California and hardy to zone 6.

Gaultheria procumbens L. BOXBERRY, CHECKERBERRY, CREEPING WINTERGREEN, IVRY-LEAVES, MIQUELBERRY, MOUNTAIN TEA, PARTRIDGEBERRY, TEABERRY, WINTERGREEN. Native from Newfoundland to Manitoba and Minnesota and south to Georgia and Alabama (zone 3), this species produces creeping stems about 12 centimeters (4.7 inches) high with dark, lustrous leaves that turn red in the autumn. They emit a strong wintergreen scent when bruised. The flowers are perfect and waxy white or pinkish white, solitary and nodding. The fruit is scarlet, hangs on the plant from summer until the following spring, and originally was used as a source for oil of wintergreen, much used as flavoring and in medicine. The oil was also distilled from the leaves but is now obtained from *Betula lenta,* a less expensive source, or is produced synthetically. The fruit at one time was common in Boston markets, and the Native Americans of Maine made a tea from the leaves (Hedrick, 1919). The species was introduced into cultivation in 1762 (Dirr, 1998). Harvest begins about two months after initiation of spring growth. Flowering occurs in summer and the fruit ripen in fall.

wolfberry: *Lycium* spp. BOXTHORN, MATRIMONY VINE, DESERT THORN, RABBIT THORN, TOMATILLA. This is a variable genus that contains many species, of which the common names are somewhat confused. All species are shrubs and many be either thorny or not, deciduous or evergreen. There are about 100 species native to temperate and tropical regions of both hemispheres. While the shrubs are mostly used as ornamentals, their bright red berries are sometimes eaten, as well as the cooked leaves of *L. chinense.* The most important species (*L. chinense* Mill. and *L. halimifolium* Mill.) are generally hardy to zone 4. *L. chinense* was introduced before 1709 while *L. halimifolium* was introduced aobut 1804.

wolly dod's rose: *See* **applerose.**

wonderberry *(Solanum melanocerasum): See* **huckleberry, garden.**

yang-tao: *See* **kiwifruit.**

yostaberry *(Ribes nigrum × R. divaricatum × R. una-crispa): See* ***Ribes.***

youngberry: EUROPEAN DEWBERRY. A hybrid of phenomenalberry × 'Austin Mayers' dewberry, also sometimes assigned to the species *Rubus caesius* L. and sometimes *R. ursinus.* The fruit of this hybrid is sweeter than that of loganberry but the plants are more cold sensitive. *See* **blackberry.**

Glossary

air layering: Also known as **marcottage**, this procedure is used often on herbaceous material but can be used on some small fruit plants. The layers can be made on dormant one-year-old wood or on current season's wood. The wood preferably should be about pencil thickness. Make a longitudinal cut two to three inches in length and deep enough to contact the center of a young shoot. Dust the cut with a hormonal rooting powder containing **IBA** or **NAA,** wrap the wounded area with about a handful of damp peat moss, and cover the moss and stem with a piece of polyethylene. The plastic should be wrapped around the stem and fastened top and bottom with twine or electrical tape. Be sure there is no opening in the plastic to allow the moss to dry out and that the tape begins on the shoot and wraps down upon the plastic to completely seal in moisture and seal out rain water. If the plastic is not sealed properly the moss may dry out *or* rain or irrigation water seeping beneath the plastic will keep the moss too soggy and reduce aeration. Either drying or overwetness will prevent roots from forming.

budding: This is a type of grafting in which a single bud is used instead of a scion. The same principles of cambium matching and coating of exposed cut surfaces with wax apply to budding as to grafting. Buds must be dormant, must be taken from current year's growth or one-year-old wood, and must be inserted right side up, as for grafting using scions.

budstick: This is a vigorous shoot about the same diameter as the stock (pencil-diameter) of the plant to be grafted that is removed from the parent plant. The budstick contains several buds, each of which in turn is removed immediately before being grafted to the stock.

cleft graft: This is often performed when the stock is considerably larger in diameter than the scion. The stock is cross-cut and then split

An Encyclopedia of Small Fruit
doi:10.1300/6102_03

along its diameter with a special grafting chisel or a hatchet, and a scion with a double-bevel at the physiological lower end is inserted into each side of the cleft so that the respective cambia can make contact. The entire junction is coated with grafting wax, and wax is applied along the split in the stock at the tops of the scions to prevent drying. In time the wound heals, the parts grow together, and shoots appear from the scion.

cuttings: A portion of a stem, a leaf, or a root that is removed from the mother plant and treated in such a way as to promote formation of roots near its physiological base. At times the addition of a hormone as a soak or a dip will aid in formation of roots. This is one of the easiest methods of propagating plants but, as with other methods, proper timing and selection of material are critical. There are several types of cuttings, but a complete discussion of cuttings is beyond the scope of this work. The author refers the reader to the many good books on plant propagation now on the market.

dichogamy: A condition in some plants wherein anthesis (pollen shedding) and ovule receptivity are not synchronized. In other words, pollen is shed before (protandry) or after (protogeny) the eggs are ready for fertilization. Since there is variation among individual trees of the same species, interplanting, especially of two or more cultivars of the same fruit, effectively overcomes dichogamy.

dioecious: This is from the Greek meaning, literally, "two hallways" or "two houses" and refers to the condition wherein the male flowers and female flowers are borne on separate plants. Asparagus, spinach, mulberries, kiwifruit, and some hollies are examples. This condition gives rise to the common expression "male plants" and "female plants." Only the female plants will produce fruit, but they must do so in the presence of an adequate pollen source (male plants). Planting one male plant to every six or ten female plants is a common means of providing for adequate pollination.

division: In crown-division, sometimes called clump-division, the process of dividing the crown of the plant into several pieces, each of which contains at least one shoot with accompanying root system. These are treated as individual plants and set out in a small nursery or directly into the field.

ellagic acid: This tannin-related compound is a putative anticarcinogen, antimutagen, and a potential inhibitor of chemically-induced
cancers (Okuda et al., 1989; Maas et al., 1991). Many small fruits
contain unusually large quantities of this compound, including raspberry, blackberry, and strawberry (Daniel et al., 1989). The compound has also been reported present in mayhaw (*Crataegus* spp.),
black currant *(Ribes nigrum),* and American cranberry *(Vaccinium
macrocarpon).* The amount of ellagic acid in the fruit varies according to the tissue sampled, the time of year, the species, and the cultivar
(Maas et al., 1992; Rommel and Wrolstad, 1993; Boyle and Hsu,
1990; Wang et al., 1994).

grafting: This is the process of attaching a piece of material from the
plant you wish to propagate (scion) to a root system (rootstock). The
junction of the two pieces, once healed, forms the "graft union."
There are several methods of grafting involved in fruit production;
those preferred for a particular species are given in the text under the
entry for that plant. No matter which graft is made, the scion must be
inserted right side up, must be dormant, and must be of wood no more
than one year old.

hardiness zones: The U.S. Department of Agriculture's hardiness
zonation has been used in this work. This is an expression of a plant's
adaptation to cold temperature, and zone numbers are based upon the
average minimum winter temperature for an area according to the following table:

Hardiness Zone	Temperature (°F)	Temperature (°C)
2b	−40 to −45	−40 to −43
3a	−35 to −40	−37 to −40
3b	−30 to −35	−34 to −37
4a	−25 to −30	−32 to −34
4b	−20 to −25	−29 to −32
5a	−15 to −20	−26 to −29
5b	−10 to −15	−23 to −26
6a	−5 to −10	−21 to −23
6b	0 to −5	−18 to −21
7a	0 to 5	−18 to −15
7b	5 to 10	−15 to −12

8a	10 to 15	−12 to −9
8b	15 to 20	−9 to −7
9a	20 to 25	−7 to −4
9b	25 to 30	−4 to −1
10a	30 to 35	−1 to 2
10b	35 to 40	2 to 4

The reader should be aware that these are average figures only and may vary considerably among winters. In addition, because specific site mesoclimates and aspects can influence temperature considerably, the designation of a specific zone is considered accurate to +/− one zone. Therefore, the south slope of a hill may experience zone 4 winter conditions while the north slope of the same hill may experience zone 3 winter conditions. In addition, the hardiness zone does not speak to variations in soil pH, salinity, texture, and water-holding capacity, atmospheric relative humidity, wind, etc. Therefore, plants adapted to the same zone may vary widely in other requirements necessary for adequate growth.

hardwood cuttings: These are about 17 to 23 centimeters (7-9 inches) long and taken from dormant, current year's or one-year-old wood. As with the selection of softwood cuttings, take hardwood cuttings from healthy, vigorous, but not overly vigorous shoots or suckers. Make the cut at the base of the shoot with a sharp knife, as cutting shears sometimes crush the wood and reduce rooting. If frozen wood is collected let it thaw in the sun before bringing it to the propagating room.

In the propagating room make the cuttings from the shoots you have collected. Make the bottom cut just below a node and the top cut just above the third node from the bottom. Wrap the cuttings in bundles of 25 and store them upside down in damp peat moss or sand in a cool place but one that remains above freezing. The bottoms will callus at this time. When the soil has warmed in spring set the cuttings outside right side up and deeply enough so that only the top inch or two remains above the soil line. Cuttings will root and can be transplanted in the autumn.

IBA: Abbreviation for the hormone indolebutyric acid. IBA stimulates root formation in cuttings and is commonly found in rooting powders. As with all hormones, concentration, and timing of application are very important. Often, too small a concentration of the hormone will have no effect on rooting, while too great a concentration may inhibit rooting. Always treat cuttings with the recommended concentration of rooting hormone.

layering: Several types of layering are used to propagate fruit plants. All contribute to the formation of roots on current year's or one-year-old growth while that growth remains attached to the mother plant, and all are easily performed with a high degree of success provided a few prerequisites are maintained. The soil temperature must be warm to promote rooting, the soil moisture high to maintain root and subsequent daughter plant growth, the portion of the shoot to be rooted wounded by nicking or scraping where it contacts the soil to promote root formation, and the portion of the shoot to be rooted be healthy and about pencil-thick.

maintenance pruning: Essentially this is the annual removal of dead, diseased, and broken wood, the removal of sufficient old wood to allow for satisfactory light penetration into the plant's canopy, the removal of unwanted and overly vigorous shoots, of water sprouts, and of unwanted suckers. Pay particular attention to the latter in the case of grafted plants, as suckers that are allowed to grow from below the graft union will do so to the detriment of the top of the plant. Maintenance pruning is practiced on established trees.

marcottage: *See* **air layering.**

mound layering: Also sometimes called stooling, this method allows for the propagation of many daughter plants in a single operation, unlike simple and tip layering, from which a single daughter plant is produced on each shoot. Healthy, established mother plants of many bushes and some single-trunk trees are cut to within a few inches of the soil in early spring. In a few weeks several adventitious shoots will arise from each stub. When these are about six inches tall, soil or sawdust is mounded over the plants so that the bottom half of the new shoots is covered, leaving the tips of the new shoots exposed. As the shoots elongate, more soil or sawdust is mounded over them, always allowing the top half of the shoots to remain exposed. In several months the shoots will have rooted, at which time the mound is removed and the daughter plants clipped from the mother plant in the autumn or in the following spring. The mother plant will then continue to send up new shoots and the process repeats over and over. At some point, depending upon the vigor of the mother plant but perhaps every couple of years, allow the mother plant to send up its shoots and grow normally so as to replenish its supply of nutrients for further daughter plant production.

NAA: Abbreviation for the synthetic hormone naphthalene acetic acid. NAA stimulates rooting in cuttings and also is used sometimes to thin fruit from heavily set trees. Always use hormones at the recommended concentration and application timing.

pedicle: The stem that attaches a flower or fruit to the main stem of the inflorescence (peduncle). The pedicle is generally smaller than the peduncle.

peduncle: The main stem of an inflorescence attached directly to the shoot of a plant.

petiole: The stem or stalk of a leaf. The petioles of rhubarb and celery are the portions of the plant that are eaten.

renewal pruning: This may take different forms on different species, but generally involves cutting old, overgrown bushes to the ground. If this is done on the entire bush in early spring, the plant may not bloom for a couple of years thereafter and the grower will be faced with having to thin out much "weedy" growth, that is, soft, succulent suckers that proliferate in such cases. Alternatively, several of the older canes can be removed each year over a period of three or four years, until all have been cut to the ground and allowed to regrow.

root cuttings: These are usually collected in early spring but in some cases can be collected in late autumn before the ground freezes and held over winter for spring planting.

Large cuttings 15 to 21 centimeters (6-8) inches long are collected in the autumn, packed in damp peat moss and stored in a cool place over the winter, then planted outdoors in early spring. Orient the cuttings so that the freshly cut ends are near the soil surface and the original basal tips are pointed down into the earth.

Cuttings made from fine roots are trimmed to 2 to 5 centimeters (1-2 inches) in length, broadcast over very fine soil in a planting container and then covered with more finely sifted soil. The container is watered, covered with polyethylene to prevent drying, and kept in a shady spot until the cuttings have rooted.

Cuttings taken from fleshy roots should be about 5 to 8 centimeters (2-3 inches) long and oriented in the medium as discussed for large cuttings. It is critically important that polarity be maintained, that is, that the cuttings be placed into the soil with the original base of the cutting pointed downward.

rooting media: Although research has shown that some species root more rapidly in specific types of media, a complete discussion of these findings is beyond the scope of this work.

Clean, sharp sand, that is, masonry sand, works well for many species. It drains easily so that the chance of rotting is diminished, but the problem becomes keeping it moist enough for rooting to occur. Peat moss is also easy to use, but since it holds moisture the problem with its use is opposite that of sand, that is, it may remain too wet and promote rot in the cuttings. A mixture of peat moss and sand is a compromise. Vermiculite, or a mixture of vermiculite and sand, also works well.

semi-hardwood cuttings: These are cuttings made from wood that is no longer soft but has not yet fully hardened and become woody. The proper time for taking these is approximately after the rapid flush of spring growth has ceased but before the plants harden for the winter, and may vary from mid to late summer in the north. They are taken in a manner similar to **hardwood cuttings** but are dipped in a rooting hormone and struck into the rooting media immediately after collection rather than being subjected to the callusing treatment.

side-veneer graft: Make a slanting cut into the side of the stock and insert the scion, the bottom of which has been shaped like a wedge, into the cut. If the wood of the stock cut has too little tension to hold the scion firmly, wrap the union with waxed string or tape to hold the pieces until they knit. Coat all exposed cut surfaces, including the top of the scion, with grafting wax. This type of graft is useful when the stock is considerably larger than the scion. The top of the stock above the graft union can be removed after the graft is made.

simple layering: A branch of one-year-old wood is bent to the soil. About six inches back of the tip the branch is wounded, covered with a few inches of soil, and held in place with a stone or a peg. A small stake is placed into the soil near the protruding tip of the branch and that tip tied to it to keep it in an upright position. In the north this is often done in the spring, and by autumn the branch has produced roots at the wounded bend. Clip off the new daughter plant and set it to the nursery or the landscape.

softwood cuttings: These require a bit more care and attention to detail than the other types of cuttings. Also as with other means of propagation, timing of cutting is of utmost importance, varying widely with the species and the location. The cutting will not root if it has been collected too late and has hardened and become woody; it will rot if collected too early and is too soft. To determine the correct time, bend the cutting. If it bends without breaking, wait a few days; if it snaps, collect it immediately.

Early in the morning, or on a cloudy day, take tip cuttings 12 to 16 centimeters (5-6 inches) long from a healthy, vigorous shoot growing in full sun to the outside of the mother plant. Avoid cuttings from diseased wood and from suckers or otherwise overly vigorous wood. Make the bottom cut just below a node. Remove all but the upper few leaves, wrap the cuttings in moist paper toweling and place them immediately in a poly bag to prevent drying. At the potting bench, dip the cuttings in a rooting hormone if indicated and place them into the rooting medium, striking the cutting into the medium up to its top leaves. Cover the rooting vessel with poly to keep the medium moist and place it in a warm spot so that the cuttings themselves are exposed to temperatures of around 21°C (70°F). If possible, supply some bottom heat in the form of heating mats so that the media remains slightly warmer than the top, or about 24° to 27°C (75-80°F). Correct moisture and temperature are critical. Too little moisture and the cuttings will dry out; too much and they will rot. Too low a temperature will slow rooting and the cuttings may rot before producing roots; too high a temperature will kill the cuttings.

stratification: Treatment of a seed under cool temperatures for a certain number of hours to overcome dormancy and promote germination. In general, seed is cleaned, dried for 24 to 48 hours, mixed with damp peat moss, and placed in a poly bag the top of which is then fastened tightly. The bag is placed in the refrigerator or other storage at 4°C (40°F), where it remains for anywhere from one to four months, depending upon the species. Seeds are then removed and sown in the usual manner.

T-budding: A T-shaped slit is made in the bark of the pencil-diameter stock and the bark carefully pulled away from the slit just enough to allow for the placement of a single bud into the slit. The bud is cut from a **budstick** just after the bark of the stock is pulled back prior to budding. The bud is cut with a shield-shaped strip of wood pointed at both ends, and with the leaf petiole intact to serve as a handle for manipulating the tiny bud. The leaf blade is removed. The bud with the wood shield is inserted into the slit in the bark of the stock. The bark is gently pulled back over the wood of the bud and the entire junction wrapped or taped and all exposed cut surfaces coated with wax. Be sure to insert the bud right-side up. Rub off the leaves immediately above and below the union and all buds of the stock above the union.

tip layering: This is quite similar to simple layering. The tip of a branch of current year's wood is inserted into the soil with no further attention, usually in mid to late summer when the season's shoot growth has ceased. Roots will form quickly and the daughter plant can be separated for planting in the spring. Some small fruit, notably gooseberry, blackberry, and black and purple raspberries, will naturally tip layer if left to their own devices.

whip, or whip and tongue, graft: This is often used when the diameters of the scion and stock are roughly equal, that is, when both are about pencil-thick. A single sloping bevel about an inch and a half in length is made in the bottom of the scion and a similar sloping cut at the top of the stock. At the remains of the pith in the center of the scion and the stock, make a cut about an inch in length and parallel to the surface of the scion and stock so as to form a notch. Both scion and stock at this point will resemble a wooden clothespin with two tails. Insert a tail of the stock into the notch of the scion and push the pieces together snugly. Lash the pieces with waxed twine or a rubber band, or wrap them with adhesive tape. Coat all exposed cut surfaces, including the cut tops of the scions, with grafting wax.

Literature Cited

Ahmedullah, M.A. 1996. Growing grapes in the home garden. In: Gough, R.E. and E.B. Poling (eds.). *Small Fruits in the Home Garden*. Binghamton, NY: Food Products Press. pp. 143-188.

Alkofahi, A., J.K. Rupprecht, J.E. Anderson, J.L. McLaughlin, K.L. Milolajczak, and B.A. Scott. 1989. Search for new pesticides from higher plants. In: Arnason, J.T., B.J.R. Philogene, and P. Morand (eds.). *Insecticides of Plant Origin*. Washington, DC: American Chemical Society Symposium Series: 2:387. pp. 25-43.

Baird, W.P. 1950. *The Home Fruit Garden on the Northern Great Plains*. USDA Farmers Bulletin 1522. Washington, DC: U.S. Department of Agriculture.

Ballington, J.R., B.W. Foushee, and F. Williams-Rutkosky. 1989. Potential of chip-budding, stub grafting or hot callusing following saddle-grafting on the production of grafted blueberry plants. *Proceedings of the Sixth North American Blueberry Research and Extension Workers Conference*. Portland, OR. pp. 114-120.

Barney, D.L. 1996. Currants, gooseberries, and jostaberries. In: Gough, R.E. and E.B. Poling (eds.). *Small Fruits in the Home Garden*. Binghamton, NY: Food Products Press. pp. 107-142.

Barney, D.L. and K. Hummer. 2005. *Currants, Gooseberries, and Jostaberries*. Binghamton, NY: Food Products Press.

Bartels, S., H. Bartholomew, M.A. Ellis, R.C. Funt, S.T. Nameth, R.L. Overmyer, H. Schneider, W.J. Twarogowski, and R.N. Williams. 1988. *Brambles: Production, Management, and Marketing*. Ohio Cooperative Extension Service Bulletin 783. Columbus, OH: Ohio State University.

Beutel, J.A. 1986. Kiwifruit training. In: *Kiwifruit Production Meeting Proceedings, Saanichton, British Columbia*. Victoria, British Columbia, Canada: Queen's Printer.

Blair, D.S. 1954. *Plum Culture*. Department of Agriculture Publication 849. Ottawa, Canada: Department of Agriculture and Agri-Food.

Boyle, J.A. and L. Hsu. 1990. Identification and quantification of ellagic acid in muscadine grape juice. *American Journal of Enology and Viticulture* 41:43-47.

Brooks, R.M. and H.P. Olmo. 1972. *Register of New Fruit and Nut Varieties*. 2nd ed. Berkeley, CA: University of California Press.

Carter, F.M. 1973. Grafting cacti. *Horticulture* 51:34-35.

Carter, P. and R.G. St. Pierre. 1996. *Growing Blueberries in Saskatchewan*. Saskatoon, Saskatchewan, Canada: Department of Horticulture Science, University of Saskatchewan–Saskatoon.

Clinton, K.K. 1998. Lycopene: Chemistry, biology, and implications for human health and disease. *Nutrition Review* 56:35-51.

Collins, J.L. 1960. *The Pineapple: History, Cultivation, Utilization*. London: Leonard-Hill.

Conner, A.M., J.J. Luby, and C.B.S. Tong. 2002. Genotypic and environmental variation in antioxidant activity, total phenolic content, and anthocyanin content among blueberry cultivars. *Journal of the American Society for Horticultural Science* 127(1):89-97.

Crandall, P.C. 1995. *Bramble Production*. Binghamton, NY: The Haworth Press, Inc.

Crandall, P.C. and H.A. Daubeny. 1990. Raspberry management. In: Galletta, G.J. and D.G. Himelrick. *Small Fruit Crop Management*. Englewood Cliffs, NJ: Prentice Hall. pp. 157-213.

Cross, C.E., I.E. Demoranville, K.H. Deubert, R.M. Devlin, J.S. Norton, W.E. Tomlinson, and B.M. Zuckerman. 1969. *Modern Cultural Practice in Cranberry Growing*. Bulletin 39. Amherst, MA: Massachusetts Agricultural Experiment Station Extension Service.

Dana, M.N. 1990. Cranberry management. In: Galletta, G.J. and D.G. Himelrick. *Small Fruit Crop Management*. Englewood Cliffs, NJ: Prentice Hall. pp. 354-362.

Daniel, E.M., A.S. Drupnick, Y.H. Heur, J.A. Blinzler, R.W. Nims, and G.D. Stoner. 1989. Extraction, stability, and quantification of ellagic acid in various fruits and nuts. *Journal of Food Chemistry and Component Analysis* 2:338-349.

Darrow, G.M. 1955. Blackberry:raspberry hybrids. *Journal of Heredity* 46:67-71.

Davis, A.R., M. Tosiano, S. Slayton, and R. Helrich. 1996. *Maine Wild Blueberries*. Concord, NH: New England Agriculture Statistics Service.

Dirr, M.A. 1998. *Manual of Woody Landscape Plants*. Champaign, IL: Stipes Publishing Co.

Dirr, M.A. and C.W. Heuser. 1987. *The Reference Manual of Woody Plant Propagation*. Athens, GA: Varsity Press.

Doncaster, T. 1981. The whys and hows of passion fruit growing in Queensland. *Combined Proceedings of the International Plant Propagation Society* 30:616-617.

Eck, P. 1988. *Blueberry Science*. New Brunswick, NJ: Rutgers University Press.

Eck, P. 1990. *The American Cranberry*. New Brunswick, NJ: Rutgers University Press.

Edinger, P. (ed.). 1970. *Succulents and Cactus*. Menlo Park, CA: Lane.

Einset, J., K. Kimball, and J. Watson. 1973. Grape varieties for New York State. *Plant Science Pomology 7*. Information Bulletin 61. Extension Publication. Ithaca, NY: New York State College of Agriculture and Life Science, Cornell University.

Emery, D.E. 1985. Arctostaphylos propagation. *Combined Proceedings of the International Plant Propagators Society* 35:281-284.

Facciola, S. 1990. *Cornucopia: A Source Book of Edible Plants*. Vista, CA: Kampong Publishing Co.

Ferguson, A.R. 1990. Kiwifruit management. In: Galletta, G.J. and D.G. Himelrick (eds.). *Small Fruit Crop Management*. Englewood Cliffs, NJ: Prentice Hall. pp. 472-503.

Flores, C.A. and C. Gallego. 1994. The production of prickly pear in the north-central region of Mexico. In: Felker, P. and J.R. Moss (eds.). *Proceedings of the Fifth Annual Texas Prickly Pear Council*. Kingsville, TX: Texas A&M University. pp. 13-28.

Fordham, I.M., B.A. Clevidence, E.R. Wiley, and R.H. Zimmerman. 2001. Fruit of autumn olive: A rich source of lycopene. *HortScience* 36(6):1136-1137.

Galletta, G.J. and R.S. Bringhurst. 1990. Strawberry management. In: Galletta, G.J. and D.G. Himelrick (eds.). *Small Fruit Crop Management*. Englewood Cliffs, NJ: Prentice Hall. pp. 83-156.

Gough, R.E. 1994. *The Highbush Blueberry and Its Management*. Binghamton, NY: Food Products Press.

Gough, R.E. 1996. Blueberries North and South. In: Gough, R.E. and E.B. Poling (eds.). *Small Fruits in the Home Garden*. Binghamton, NY: Food Products Press. pp. 71-106.

Gough, R.E. 1997. *Smart Gardeners Guide to Growing Fruit*. Mechanicsburg, PA: Stackpole Books.

Gough, R.E. 2002. *Juneberries for Montana Gardens*. MontGuide 198806. Bozeman, MT: Montana Extension Service.

Hall, I.V. 1969. *Growing Cranberries*. Canada Department of Agriculture Publication 1282 (rev.). Ottawa, Canada: Canada Department of Agriculture.

Hansen, N.E. 1940. *New Hardy Fruits for the Northwest*. Bulletin 339. Brookings, SD: South Dakota Agriculture Experiment Station.

Hansen, O.B. 1990. Propagating *Sorbus aucuparia* L. and *Sorbus hybrida* L. by softwood cuttings. *Scientia Horticulturae* 42:169-175.

Harmat, L., A. Porpaczy, D.G. Himelrick, and G.J. Galletta. 1990. Currant and gooseberry management. In: Galletta, G.J. and D.G. Himelrick (eds.). *Small Fruit Crop Management*. Englewood Cliffs, NJ: Prentice Hall. pp. 245-272.

Harris, R.E. 1972. *The Saskatoon*. Publication 1246. Ottawa, Canada: Canada Department of Agriculture.

Hartmann, H.T., D.E. Kester, F.T. Davies, and R.L. Geneve. 2002. *Plant Propagation*. 7th ed. Englewood Cliffs, NJ: Prentice Hall.

Hedrick, U.P. 1919. *Sturtevant's Notes on Edible Plants*. Twenty-Seventh Annual Report of the New York Agricultural Experiment Station. Volume 2, Part 2. Albany, NY: J.P. Lyon Co.

Hilton, R.J. 1982. Registration of *Amelanchier* cultivar names. *Fruit Varieties Journal* 36(4):108-110.

Hopping, M.E. 1986. Kiwifruit. In: Monselise, S.P. (ed.). *CRC Handbook of Fruit Set and Development*. Boca Raton, FL: CRC Press. pp. 217-232.

Hortus Third: A Concise Dictionary of Plants Cultivated in the United States and Canada. 1976. Cornell University. New York: Macmillan Company.

Howard, G.S. and G.B. Brown. 1962. *Hardy, Productive Tree Fruits of the High Altitude Section of the Central Great Plains.* USDA-ARS 34-40. Washington, DC: U.S. Department of Agriculture.

http://www.faostat.fao.org. Retrieved December 2005.

http://www.fas.usda.gov/psd/complete_tables/HTP. Retrieved December 2005.

http://www.jan.mannlib.cornell.edu/reports/massr/fruit/pmf-bb/meit0105.txt. Retrieved December 2005.

Hummer, K. 1995. The mystical powers and culinary delights of the hazelnut: A globally important Mediterranean crop. *Diversity* 11(1/2):130.

Hummer, K. 1999. Corylus genetic resources. *Hazelnut.* USDA ARS National Clonal Germplasm Repository, Corvallis, OR. Washington, DC: U.S. Department of Agriculture. Also available online: http://www.ars.usda.gov/main/docs .htm?docid=11035.

IBPGR. 1986. *Genetic Resources of Tropical and Subtropical Fruits and Nuts.* Rome, Italy: International Board of Plant Genetics Resources.

Jacob, H.E. 1959. *Grape Growing in California.* Circular 116. Berkeley, CA: California Agricultural Extension Service, University of California.

Jaffee, A. 1970. Chip grafting guava cultivars. *The Plant Propagator* 16(2):6.

Jay, D. and C. Jay. 1984. Observations of honeybees on Chinese gooseberries ("kiwifruit") in New Zealand. *Bee World* 65(4):155-166.

Jones, G.N. 1946. *American Species of* Amelanchier. Urbana, IL: University of Illinois Press.

Knight, R. 1980. Origin and world importance of tropical and subtropical fruit crops. In: S. Nagy and P.E. Shaw (eds.). *Tropical and Subtropical Fruits: Composition, Properties, and Uses.* Westport CT: AVI Publishing Co. pp. 1-120.

Kohlmeier, L., J.D. Kark, E. Gomez-Garcia, B.C. Martin, S.E. Steck, A.F.M. Kardinaal, J. Ringstad, M. Thamm, V. Masaev, R. Riemersma, J.M. Martin-Moreno, J.K. Huttunen, and F.J. Kok. 1997. Lycopene and myocardial infarction risk in the EURAMIC study. *American Journal of Epidemiology* 146:618-626.

Kumar, G.N. Mohan. 1990. Pomegranate. In: S. Nagy, P.E. Shaw, and W.F. Wardowski (eds.). *Fruits of Tropical and Subtropical Origin.* Longboat Key, FL: Florida Science Source. pp. 328-347.

Kunde, R.M., K.A. Lider, and R.V. Schmidt. 1968. A test of *Vitis* resistance to *Xiphinema index. American Journal of Enology and Viticulture* 19:30-36.

LaRue, J. 1980. *Growing pomegranates in California.* DANR Pub. Leaflet 2459. Berkeley, CA: University of California, Division of Agriculture Sciences.

Laughlin, K.M., R.C. Smith, and R.G. Askew. 1988. *Juneberry for Commercial and Home Use on the Northern Great Plains.* Published jointly as Bozeman, MT: Montana State University Extension Service MT 8806 and Fargo, ND: North Dakota State University Extension Service H-938.

Logan, M. 1996. *The Packer: Produce Availability and Merchandising Guide.* Lincolnshire, IL: Vance Publishing Corp.

Lutz, J.M. and R.E. Hardenburg. 1968. *The Commercial Storage of Fruits, Vegetables, and Florist and Nursery Stocks.* Agriculture Handbook 66. Washington, DC: U.S. Department of Agriculture.

Maas, J.L., G.J. Galletta, and G.D. Stoner. 1991. Ellagic acid, an anticarcinogen in fruits, especially in strawberries: A review. *HortScience* 26:10-14.

Maas, J.L., G.J. Galletta, and S.Y. Wang. 1992. Ellagic acid enhancement in strawberries. pp. 345-362. In: D.D. Bills and S.D. Kung (eds.). *Biotechnology and Nutrition: Proceedings of the 3rd Symposium.* Oxford, UK: Butterworth-Heineman.

Majumdar, P.K and S. K Mukherjee. 1968. Guava: A new vegetative propagation method. *Indian Horticulture* 12(2): 11-35.

Markle, G.M,, J.J. Baron, and B.A. Schneider. 1998. *Food and Feed Crops of the United States.* 2nd ed. Euclid, OH: Meister Publishing Co.

Miller, W.S. and C. Stushnoff. 1971. A description of *Amelanchier* species in regard to cultivar development. *Fruit Varieties and Horticulture Digest* 25(1):3-10.

Moore, J.N. and D.P. Ink. 1964. Effect of rooting medium, shading, type of cutting, and cold storage of cuttings on the propagation of highbush blueberry varieties. *Proceedings of the American Society for Horticultural Science* 85:285-294.

Moore, J.N. and R.M. Skirvin. 1990. Blackberry management. In: Galletta, G.J. and D.G. Himelrick (eds.). *Small Fruit Crop Management.* Englewood Cliffs, NJ: Prentice Hall. pp. 214-244.

Morton, J.F. 1987. *Fruits of Warm Climates.* Greensboro, NC: Media, Inc.

Mowry, H., L.R. Toy, and H.S. Wolfe. 1958. *Miscellaneous Tropical and Subtropical Florida Fruits.* Florida Extension Bulletin 156A. Gainesville, FL: Florida Extension Service.

Okuda, T., T. Yoshida, and T. Hatano. 1989. Ellagitannins as active constituents of medical plants. *Planta Medica* 55:117-122.

Olsen, J. Undated. *Nut Growers Handbook.* Oregon State University Extension Service. Corvallis, OR: Oregon Extension Service.

Ourecky, D.K. 1977. The elderberry, an indigenous American fruit. *The Cornell Plantations* 33(1):7-11.

Patterson, C.F. 1957. *Fruit Gardening in Saskatchewan.* Agriculture Extension Bulletin 123. Saskatoon, Saskatchewan: University of Saskatchewan, College of Agriculture.

Peker, K. 1962. Human health and the filbert, a medical history. *Peanut Journal and Nut World* 41(5):30-31, 38-40.

Pennock, W. and G. Maldonado. 1963. The propagation of guavas from stem cuttings. *Journal of Agriculture University of Puerto Rico* 47:280-290.

Poling, E.B. 1996. Blackberries. In: Gough, R.E. and E.B. Poling (eds.). *Small Fruits in the Home Garden.* Binghamton, NY: Food Products Press. pp. 33-71.

Poling, E.B. 1996. Strawberries for the home garden. In: Gough, R.E. and E.B. Poling (eds.). *Small Fruits in the Home Garden.* Binghamton, NY: Food Products Press. pp. 227-258.

Prescott, J.A. 1965. The climatology of the vine (*Vitis vinifera* L.) and the cool limits of cultivation. *Transactions of the Royal Society of South Australia* 89:5-23.

Pritts, M.P. 1996. Raspberries. In: Gough, R.E. and E.B. Poling (eds.). *Small Fruit in the Home Garden*. Binghamton, NY: Food Products Press. pp. 189-226.

Rehder, A. 1947. *Manual of Cultivated Trees and Shrubs*. New York: The Macmillan Co.

Remlinger, B. and R.G. St. Pierre. 1995. *Biology and Culture of the Buffaloberry*. Saskatoon, Saskatchewan, Canada: Department of Horticulture Science, University of Saskatchewan–Saskatoon.

Retamal, N., J.M. Duran, and J. Fernandez. 1987. Seasonal variations of chemical composition in prickly pear *(Opuntia ficus-indicus)*. *Journal of Scientific Food Agriculture* 38:303-311.

Ritter, C.M. and G.W. McKee. 1964. *The Elderberry: History, Classification, and Culture*. Bulletin 709. University Park, PA: Pennsylvania State University Agricultural Experiment Station.

Rommel, A. and R.E. Wrolstad. 1993. Ellagic acid content of red raspberry juice as influenced by cultivar, processing, and environmental factors. *Journal of Agriculture and Food Chemistry* 14:1951-1960.

Rupprecht, J.K., C.J. Chang, J.M. Cassady, and J.L. McLaughlin. 1986. Asimicin, a new cytotoxic and pesticidal acetogenin from the pawpaw, *Asiminia triloba* (Annonaceae). *Heterocycles* 24:1197-1201.

Sander, E. 1963. The filbert tree and how to plant it. *Nut Growers Society of Oregon and Washington* 49:43-44.

Schreiber, A. and L. Ritchis. 1995. *Washington Minor Crops*. Food and Environmental Quality Lab. Pullman, WA: Washington State University.

Shaulis, N., J. Einset, and A.B. Pack. 1968. *Growing Cold Tender Grape Varieties in New York*. New York State Agricultural Experiment Station Bulletin 821. Geneva, NY: Cornell University.

Smith, G.S., C.J. Asher, and C.J. Clark. 1987. *Kiwifruit Nutrition: Diagnosis of Nutrition Disorders*. 2nd rev. ed. Wellington, New Zealand: Agpress Communications.

Snyder, E. 1937. Grape development and improvement. In: *Yearbook of Agriculture*. Washington, DC: U.S. Department of Agriculture. pp. 631-664.

Stang, E.J. 1990. Elderberry, highbush cranberry, and juneberry management. In: Galletta, G.J. and D.G. Himelrick (eds.). *Small Fruit Crop Management*. Englewood Cliffs, NJ: Prentice Hall. pp. 363-382.

Stang, E.J., B.A. Birrenkott, and J. Klueh. 1993. Response of "Erntedank" and "Koralle" lingonberry to preplant soil organic matter incorporation. *Journal of Small Fruit and Viticulture* 2(1):3-10.

Stephens, J.M. 1988. *Manual of Minor Vegetables*. Florida Cooperative Extension Service Bulletin SP-40. Gainesville, FL: University of Florida.

St. Pierre, R.G. 1993. *The Chokecherry: A Guide for Growers*. Saskatoon, Saskatchewan, Canada: Department of Horticulture Science, University of Saskatchewan–Saskatoon.

St. Pierre, R.G. 1996. *The Lingonberry*. Saskatoon, Saskatchewan, Canada: Department of Horticulture Science, University of Saskatchewan–Saskatoon.

Teulon, J. 1971. Propagation of passion fruit *(Passiflora edulis)* on a fusarium-resistant rootstock. *Plant Propagator* 17(3):4-5.

Tomkins, J.P. 1977. Cane and bush fruits are the berries: Often it's grow them or go without. pp. 272-278. In: *Growing Fruits and Nuts*. Agricultural Information Bulletin No. 408. Washington, DC: U.S. Department of Agriculture.

Tous, J. and L. Ferguson. 1996. Mediterranean fruits. In: Janick, J. (ed). *Progress in New Crops*. Arlington, VA: ASHS Press. pp. 416-430.

Tulloch, H. and R.G. St. Pierre. 1996. *Growing Black Currants in Saskatchewan*. Saskatoon, Saskatchewan, Canada: Department of Horticulture Science, University of Saskatchewan–Saskatoon.

Van Dyk, M. and R. Currah. 1983. Vegetative propagation of prairie forbs native to southern Alberta, Canada. *The Plant Propagator* 28:12-14.

Wallis, J.S. 1976. The propagation and training of standard fuchsias. *Combined Proceedings of the International Plant Propagation Society* 26:346-348.

Wang, S.Y., J.L. Maas, J.A. Payne, G.J. Galletta. 1994. Ellagic acid content in small fruits, mayhaws, and other plants. *Journal of Small Fruit and Viticulture* 2(4): 39-50.

Wasley, R.1979. The propagation of *Berberis* by cuttings. *Combined Proceedings of the International Plant Propagators Society* 29:215-216.

Weaver, R.J. 1976. *Grape Growing*. New York: John Wiley & Sons.

Winkler, A.J., J.A. Cook, W.M. Kliewer, and L.A. Lider. 1974. *General Viticulture*. Berkeley, CA: University of California Press.

Woodroof, J.G. 1979. *Tree Nuts*. Westport, CT: AVI Publishing Co., Inc.

Wyman, D. 1969. *Shrubs and Vines for American Gardens*. New York: The Macmillan Co.

Wyman, D. 1986. *Wyman's Gardening Encyclopedia*. New York: The Macmillan Co.

Yeager, A.F. 1935. *Growing Fruit in North Dakota*. Bulletin 280. Fargo, ND: North Dakota Agricultural Experiment Station.

Zieslin, N, and A. Keren. 1980. Effects of rootstock on cactus grafted with an adhesive. *HortScience* 21:153-154.

Index

Acerola. *See* Cherry, Barbados
Actinidia arguta. See Kiwifruit
Actinidia, bower (*Actinidia arguta*).
    See Kiwifruit
Actinidia chinensis. See Kiwifruit
Actinidia deliciosa. See Kiwifruit
Actinidia kolomikta. See Kiwifruit
Actinidia polygama. See Kiwifruit
Afghanistan, pomegranate, 85
Africa, Hottentot fig, 39
Akebia, 3
Alkekengi (*Physalis alkekengi*). *See*
    Cherry, ground
Almond, flowering (*Prunus japonica*).
    See Prunus
Alpine cranberry. *See* Lingonberry
Amatungula. *See* Carissa
Amelanchier alnifolia. See Serviceberry
Amelanchier asiatica. See Serviceberry
Amelanchier canadensis. See
    Serviceberry
Amelanchier florida. See Serviceberry
Amelanchier oblongifolia. See
    Serviceberry
Amelanchier ovalis. See Serviceberry
Amelanchier sanguinea. See
    Serviceberry
Amelanchier spicata. See Serviceberry
Amelanchier stolonifera. See
    Serviceberry
Antidesma bunius. See Bignay
Apple
    baked (*Rubus chamaemorus*). *See*
        Blackberry
    belle (*Passiflora laurifolia*). *See*
        Passionfruit
    conch (*Passiflora maliformis*)

Appleberry, 4
Applerose, 4-5
Apricot
    Manchurian bush (*Prunus*
        armeniaca var. *mandshurica*).
        See Prunus
    Siberian bush (*Prunus armeniaca*
        var. *siberica*). *See* Prunus
Arctostaphylos uva-ursi. See Bearberry
Arctostaphylos. See Bearberry
Arkansas, ribes (currants), 104
Arm, grape vine, 50
Aronia arbutifolia. See Chokeberry
Aronia melanocarpa. See Chokeberry
Aronia prunifolia. See Chokeberry
Ash, mountain, 5
Asia
    autumn olive, 75
    barberry, 6-7
    governor's plum, 43
    kiwifruit, 63, 64
    whortleberry, 126
Asimina. See Banana, custard
Asimina triloba. See Banana, custard
Australia
    appleberry, 4
    bignay, 8-9
    hottentot fig, 39
    passionfruit, 76
    prickly pear, 80
Autumn elaeagnus. *See* Olive, autumn
Avellana (*Corylus avellana*). *See*
    Filbert

Baked apple berry. *See* Blackberry
Bahamas, Cherry, Barbados, 24

147

Chinese quince. *See* Quince, flowering
Chinook, bearberry, 8
Chippewa, bearberry, 8
Chocolate vine (*Akebia quinata*). *See* Akebia
Chokeberry, 27-28
Chokecherry (*Prunus virginiana*). *See* Prunus
Chuckley-plum (*Prunus virginiana*). *See* Prunus
Cloudberry (*Rubus chamaemorus*). *See* Blackberry
Columbia manzanita. *See* Bearberry
Conch nut. *See* Passionfruit
Coryberry. *See* Blackberry
Corylus americana. *See* Filbert
Corylus avellana. *See* Filbert
Corylus cornuta. *See* Filbert
Corylus heterophylla. *See* Filbert
Corylus maxima. *See* Filbert
Costa Rica
    blackberry, 9-12
    pineapple, 82
Cowberry. *See* Cranberry, highbush
Crakeberry (*Empetrum nigrum*). *See* Crowberry
Crampbark (*Viburnum opulus*). *See* Cranberry, highbush
Cranberry, 28-30
    Alpine (*Vaccinium vitis-idea*). *See* Lingonberry
    dry ground (*Vaccinium vitis-idea*). *See* Lingonberry
    highbush, 30-32
    hog (*Arctostaphylos uva-ursi*). *See* Bearberry
    lowbrush (*Vaccinium vitis-idea*). *See* Lingonberry
    moss (*Vaccinium vitis-idea*). *See* Lingonberry
    mountain (*Vaccinium vitis-idea*). *See* Lingonberry
    rock (*Vaccinium vitis-idea*). *See* Lingonberry
Cranberry tree. *See* Cranberry, highbush

Crataegus. *See* Hawthorn
Creashak (*Arctostaphylos uva-ursi*). *See* Bearberry
Cree, bearberry, 8
Creeping barberry. *See* Grape, Oregon
Creeping mahonia. *See* Grape, Oregon
Crimsonberry (*Rubus arcticus*). *See* Blackberry
Crowberry, 32
Crucifixion berry (*Shepherdia argentea*). *See* Buffaloberry
Cup, sweet (*Passiflora maliformis*). *See* Passionfruit
Curlewberry (*Empetrum nigrum*). *See* Crowberry
Currant. *See* Ribes
Currant tree. *See* Serviceberry
Curuba (*Passiflora mollissima*). *See* Passionfruit

Dangleberry (*Gaylussacia frondosa*). *See* Huckleberry
Date
    Chinese (*Ziziphus jujuba*). *See* Jujube
    trebizond. *See* Olive, Russian
Deerberry. *See* Blueberry
Delaware, Cranberry, 29
Desert thorn. *See* Wolfberry
Devil's-shoestrings (*Viburnum alnifolium*). *See* Cranberry, highbush
Dewberry (*Rubus macropetalus*). *See* Blackberry
Dog-hobble (*Viburnum alnifolium*). *See* Cranberry, highbush
Dogberry. *See* Ash, mountain; Cranberry, highbush
Dogwood, white mountain (*Viburnum alnifolium*). *See* Cranberry, highbush
Dominican Republic, pineapple, 82
Dovyalis abyssinica, 33
Drupe fruit, 87

Maules quince. *See* Quince, flowering

Maypop (*Passiflora incarnata*). *See*
  Passionfruit

Mealberry (*Arctostaphylus uva-ursi*).
  See Bearberry

Mespilus, snowy. *See* Serviceberry

Mexican barberry. *See* Grape, Oregon

Mexican breadfruit. *See* Bearberry

Mexico
  cherry, ground (*Physalis*), 26
  monstera, 73
  passionfruit, 77
  pineapple, 81
  pitaya, 84
  pomegranate, 85
  prickly pear, 79, 80
  strawberry, 118

Michigan
  cranberry, 29
  strawberry, 117

Michigan banana. *See* Banana, custard

Miltomate (*Physalis ixocarpa*). *See*
  Cherry, ground

Minnesota
  cranberry, 29
  prunus, 92
  ribes (currants), 104
  silverberry, 116

Miquelberry. *See* Wintergreen

Missey-moosey (Sorbus americana).
  See Ash, mountain

Molka (*Rubus chamaemorus*). *See*
  Blackberry

Mongolian cherry, (*Prunus tomentosa*).
  See Prunus

Monox. *See* Crowberry

Monstera (*Monstera deliciosa*) , 73

Montana
  blue huckleberry. *See* Blueberry
  elderberry, 34
  prickly pear, 79

Montenegro, raspberry, 100

Mooreberry (*Vaccinium uliginosum*).
  See Blueberry

Moose bush (*Viburnum alnifolium*). *See*
  Cranberry, highbush

Moosewood-hopple (*Viburnum
  alnifolium*). *See* Cranberry,
  highbush

Morel, petty (*Solanum
  melanocerasum*). *See*
  Huckleberry, garden

Mortinia (*Vaccinium mortinia*). *See*
  Blueberry

Moss cranberry. *See* Lingonberry

Mountain ash, 5

Mountain box. *See* Bearberry

Mountain cranberry. *See* Lingonberry

Mountain juneberry. *See* Serviceberry

Nanking cherry (*Prunus tomentosa*).
  See Prunus

Nanny plum. *See* Cranberry, highbush

Nannyberry. *See* Cranberry, highbush,
  Viburnum, blackhaw

Natal-plum. *See* Carissa

Native Americans
  bearberry, 8
  prickly pear, 79
  wintergreen, 127

Native cherry. *See* Cherry, Barbados

Nebraska banana. *See* Banana, custard

Nebraska currant. *See* Buffaloberry

Nectarberry, 74. *See also* Blackberry

Nessberry, 74. *See also* Blackberry

Nevada bearberry, 8

New Hampshire
  cranberry, 29
  highbush cranberry, 31

New Mexico
  elderberry, 34
  prickly pear, 79

New York
  banana, custard, 5-6
  cranberry, 29
  grape, 45
  strawberry, 117

New Zealand
  kiwifruit, 63, 64
  passionfruit, 76, 77

Rubus coronarius. See Blackberry,
    Cranberry, highbush
Rubus cuneifolius. See Blackberry
Rubus deliciosus. See Raspberry
Rubus dumetorium. See Blackberry
Rubus flagellaris. See Blackberry
Rubus hispidus. See Blackberry
Rubus idaeus. See Raspberry
Rubus laciniatus. See Blackberry
Rubus macropetalus. See Blackberry
Rubus neglectus. See Raspberry
Rubus niveus. See Raspberry
Rubus occidentalis. See Raspberry
Rubus parviflorus. See Raspberry
Rubus phoenicolasius. See Raspberry
Rubus rosifolius. See Raspberry
Rubus spectabilis. See Raspberry
Rubus ulmifolius. See Blackberry
Rubus ursinus. See Blackberry,
    Boysenberry
Rubus vitifolius. See Blackberry
Running pop. *See* Passionfruit
Russia
    buckthorn, sea, 19
    pomegranate, 85
    strawberry, 117, 118
    raspberry, 100
    ribes (currants), 105

Sallow thorn (*Hippophae rhamnoides*).
    See Buckthorn, sea
Salmonberry (*Rubus parviflorus*). *See*
    Raspberry
Sambucus caerulea. See Elderberry
Sambucus callicarpa. See Elderberry
Sambucus canadensis. See Elderberry
Sambucus ebulis. See Elderberry
Sambucus melancarpa. See Elderberry
Sambucus nigra. See Elderberry
Sambucus pubens. See Elderberry
Sambucus racemosa. See Elderberry
Sand plum. *See* Guava
Sandberry (*Arctostaphylos uva-ursi*).
    See Bearberry
Sarviceberry. *See* Serviceberry

Sarvis. *See* Serviceberry
Saskatoon (*Amelanchier alnifolia*). *See*
    Serviceberry
Scuppernong (*Vitis rotundifolia*). *See*
    Grape
Sea buckthorn, 19
Sea tomato. *See* Rosehips
Serbia, raspberry, 100
Service tree, wild (*Sorbus torminalis*).
    See Chequers
Serviceberry (*Amelanchier*), 111-115
Serviceberry (*Sorbus americana*). *See*
    Ash, mountain
Shad. *See* Serviceberry
Shadberry. *See* Serviceberry
Shadblow. *See* Serviceberry
Shadbush. *See* Serviceberry
Sheepberry, 116. *See also* Cranberry,
    highbush
Shepherdia argentea. See Buffaloberry
Shepherdia. See Buffaloberry
Shot huckleberry. *See* Blueberry
Siberia, Kiwifruit, 63
Siberian bush apricot. *See* Prunus
Silverberry, 116. *See also* Olive,
    autumn; Olive, Russian
Six-arm kniffen, vine training system,
    49
Skunkbush (*Rhus trilobata* var.
    malacophylla). *See*
    squawbush
Sloe (*Prunus americana*). *See* Prunus
Snowball tree. *See* Cranberry, highbush
Snowy mespilus. *See* Serviceberry
Solanberry (*Solanum melanocerasum*).
    See Huckleberry, garden
Solanum melanocerasum. See
    Huckleberry, garden
Sorbus americana. See Ash, mountain
Sorbus torminalis. See Wintergreen
Sourtop (*Vaccinium myrtilloides*). *See*
    Blueberry, Whortleberry
South Africa
    kei-Apple (*Dovyalis caffra*), 62
    passionfruit, 76
    prickly pear, 80